不止设计

BEYOND DESIGN

设计

郭晰纹 著

北方联合出版传媒（集团）股份有限公司
辽宁科学技术出版社

图书在版编目（CIP）数据

不止设计 / 郭晰纹著 . — 沈阳 : 辽宁科学技术出版
社 , 2024.6
　　ISBN 978-7-5591-2957-4

　　Ⅰ . ①不… Ⅱ . ①郭… Ⅲ . ①建筑设计—作品集—中
国—现代 Ⅳ . ① TU206

　　中国国家版本馆 CIP 数据核字 (2023) 第 053537 号

出版发行：辽宁科学技术出版社
　　　　　（地址：沈阳市和平区十一纬路 25 号　邮编：110003）
印　刷　者：广东省博罗县园洲勤达印务有限公司
经　销　者：各地新华书店
幅面尺寸：230mm×210mm
印　　张：17.2
字　　数：600 千字
出版时间：2024 年 6 月第 1 版
印刷时间：2024 年 6 月第 1 次印刷
责任编辑：杜丙旭　郭芷夷
特约编辑：吴耀隆　陈柯翡
封面设计：高　斌
版式设计：郭芷夷
责任校对：王玉宝　杨　键

书　　号：ISBN 978-7-5591-2957-4
定　　价：298.00 元

编辑电话：024-23280272
邮购热线：024-23284502

每一个人就是一个"不止"的世界。因为"不止"，所以多彩，所以深刻，所以无穷。艺术家的"不止"，折射出的是她灵魂的光芒……

——何建明　著名作家

这是一本有野心的书。设计是需要有野心的，所谓有野心的设计，在于不仅能够将身体、情感与环境相关联，而且能抽离思想，将无差别的空间（space）变为有意义的地方（place）。

——杨溟　著名跨界研究者

设计如同人生，不同的认知阶段，对同一事物会有不同的表达形式，就像佛教禅宗中的从见山为山，到见山不是山，再到见山为山一样，晰纹女士的设计已经从设计到达了不止设计的意境，能与文化相融而又不止于设计，大道无形！有灵魂的设计才是精彩的设计！

——范文浩　著名学者

众所周知，设计师这个职业并不是所有人都能胜任的，除了需要具备独特的创意和审美观念，还要能够从不同的角度和层面去思考和解决问题，在我看来，最宝贵的就在于持之以恒。我体会郭晰纹提出的不止设计，其实就是在设计之外还有更多的延展空间，更深刻的人生体会，以及更繁杂的义务和责任心。

——吴耀隆　著名建筑师

强烈推荐，没有理由！

——邹学俊　亦师亦友

Every person is a world that goes beyond himself. "Beyond" means being colorful, profound, and infinite. "Beyond design", for an artist, reflects the radiance of her soul...

—Jianming He　Famous Writer

This is an ambitious book. Design requires ambition. The so-called ambitious design should be able to not only connect the body, emotions, and environment, but to detach ideas and turn indistinguishable spaces into meaningful places.

—Ming Yang　Renowned Transboundary Researcher

Design is like life, and at different stages of cognition, there will be different forms of expression for the same thing. Just like in Zen, where seeing a mountain is deemed as going beyond the mountain, Xiwen Guo's design has gone beyond design, blending with culture and reaching an intangible realm. Only designs with a soul are wonderful designs!

—Wenhao Fan　Prominent Scholar

As is well known, the seemingly glamorous profession of a designer is not something that everyone can handle. In addition to possessing unique creativity and aesthetic concepts, and being able to think and solve problems from different perspectives and levels, in my opinion, the most valuable thing is perseverance. In my understanding, Xiwen Guo's concept of "beyond design" is about having more room for the extension of design, about a deeper understanding of life, and a more complex sense of mission about obligation and responsibility.

—Yaolong Wu　Famous Architect

Strongly recommended, no reason!

—Xuejun Zhou　Mentor and Friend

郭晰纹

晰纹设计　创始人、总设计师
中国建筑师原创跨界家居品牌宇宙奇迹　联合创始人
米兰理工大学　设计管理硕士
上海青年设计领袖
中国百名优秀室内建筑师
中国文化部创意产业人才库成员
中国装饰协会十大杰出青年提名
江苏省室内设计学会常务理事
江苏省优秀青年设计师
注册高级室内建筑师

80后设计师郭晰纹女士出生于中国四川，曾先后在南京、上海、米兰修读设计，现居上海。

晰纹女士于2005年创立晰纹设计，一个注重深刻理解和创新思考空间功能的室内设计品牌。开创"室内复合空间"设计理念，以去风格化的设计语言，诠释每个空间的独特气场。在她的带领下，其团队多次荣获国内外知名奖项，包括CIID中国室内设计奖、A'DESIGN、DNA Paris Design Awards、International Design Awards、German Design Award……

她认为，设计，是对生命的阅读。设计之外，晰纹女士对生活的热爱，她的茶，她的酒，她的娃和她的狗……她的创造力以及对生命的珍惜都反映在她独特的设计作品中：复合但不混乱，大胆而不极端，美而有用，强大而克制。设计但不设限。

2022年，晰纹女士与先生建筑师吴耀隆、好友平面设计师高斌女士和美学生活家康康女士共同创立了美学集合品牌"宇宙奇迹"，以原创家居、软装和珠宝饰品等产品为核心，而品牌名称则来源于其女儿的奇思妙想。同年，精心打造了宇宙奇迹生活艺术馆，旨在实现生活与艺术的完美融合，创造一个开放包容且富有艺术氛围的场所。

Xiwen Guo

Founder of XY + Z DESIGN, chief designer

Founder of XY+Z, a cross-border home furnishings brand

Master's degree in Design Management from Politecnico di Milano

Shanghai Youth Design Leader

Top 100 Excellent Interior Architects in China

Member of the Creative Industry Talent Pool of the Ministry of Culture of China

Nomination for Top Ten Outstanding Youth of China Decoration Association

Executive Director of Jiangsu Interior Design Society

Excellent Young Designer in Jiangsu Province

Registered Senior Interior Architect

Ms. Xiwen Guo is a designer born in the 1980s in Sichuan Province, China. She studied design in Nanjing, Shanghai, and Milan, and currently resides in Shanghai.

Xiwen Guo founded XY + Z DESIGN in 2005, an interior design office that emphasizes profound understanding and innovative thinking of space and functions. She created the concept of "indoor composite space", trying to interpret the unique atmosphere of each space with a de-stylized design language. Under her leadership, her team has won several well-known domestic and international awards, including CIID China Interior Award, A'DESIGN, DNA Paris Design Awards, International Design Awards, and German Design Awards.

She believes that design is like reading life. Beyond design, Ms. Guo loves life, tea, wine, her children, and her dog… Her creativity and appreciation for life are reflected in her unique design: composite yet not chaotic, bold yet not extreme, beautiful yet useful, powerful yet restrained. Design, without limitations.

In 2022, Ms. Guo, along with her husband architect Yaolong Wu, her friends Ms. Bin Gao (a graphic designer) and Ms. Kang Kang (a lifestyle expert), jointly founded the cross-border brand "XY+Z", which focuses on original home furnishings, soft decoration, and jewelry products. The brand name is derived from her daughter's creative idea. In the same year, the XY+Z GALLERY was established with the aim of achieving a perfect integration of life and art, creating a gallery with an open, inclusive, and artistic atmosphere.

目录

CONTENTS

自序
PREFACE

似乎这本书从开始计划到最终面市，经历了太多年。原因也许与书名相关——"不止"。无论从未出过书的迷茫无措，还是抽不出时间整理完善，以及太多不满意带来的反复调整……好在，完成了。

大概每个女孩小时候都有很多关于美的梦想。我也不例外。

服装和珠宝设计师、画家，这些都是我儿时的梦想。

中学后的日子，开始觉得"家"是这个世界上最温暖的地方。能为无数人创造无数个温暖的家，该会是多么有成就感的事情。于是，我放弃了成为服装和珠宝设计师、画家的梦想。2005 年，在南京创立了最初 20 平方米的小小"熙文装饰"，开启了我"造暖"的室内设计之路。经过 5 年的设计施工经验累积，2010 年"取消"施工，专注于设计。2014 年，随先生吴耀隆一起在上海成立了"晰纹设计"，又一步步走到今天。一路走来，回首个中滋味，不禁感慨万千。

生活中的我有着多重身份：我母亲的女儿，我女儿的母亲，我先生的夫人，我同事的同事……常常在想，既然人的身份可以有这么多重，为什么空间不可以？客厅也可以是书房，餐厅不一定就只是吃饭的地方。我深信，设计的魅力在于其无限的可能性。而我所倡导的复合空间设计理念，便是希望打破传统的空间界限，让每一个空间都拥有多重身份与功能，并如有机生命体般自然生长。再后来，或许还是对儿时的梦想念念不忘。时常问自己：我能不能既做室内设计，又做服装、珠宝设计呢？答案是肯定的。就这样，2021 年，由女儿吴若可命名的"宇宙奇迹"跨界产品公司诞生。她集"珠宝、服饰、家居、艺术品"于一体。这一年，我的梦想，实现了。

在这本《不止设计》中，我真实记录了自己 18 年的设计生涯，也分享了许多关于设计、关于生活的思考与感悟，算是对即将走完的前半生的交代——无论对自己，还是对与我有关的人或事或物。于是，便有了您现在手中的这本《不止设计》。初次写书，不足之处，还请指正。

至于为 100 位平凡而又不平凡的人策划 100 场个展，也就是百场个展计划，将会是我人生下半场的梦想。希望借由个展的形式，将自己的设计理念、创作过程以及对于美的独特理解，呈现给更多的人。每一场展览都将是一个全新的开始，是我与被展人、与观众共同探索美、探索生命的旅程。

最后，感谢我的父母，给予我生命；感谢我的客户，给予我成长；感谢我的亲人、朋友，给予我支持与信任；感谢我的合伙人、同事，陪我风雨同路；感谢支持我筹备这本书的老师、同事，让它得以和大家见面；感谢设计，让我的人生，得以精彩。

郭晰纹

2024 年春 于上海

It seems that this book has gone through too long, from planning to finally hitting shelves. Perhaps it is related to the title "beyong design". From the confusion of having never published a book, to not being able to find the time to sort it out, and the repeated adjustments due to dissatisfaction... Fortunately, it is finally completed.

I think every girl has many dreams about beauty when she was a child, and I am no exception.

Fashion and jewelry designer, painter... these were my childhood dreams.

After high school, I began to feel that "home" is the warmest place in the world. It would be a very fulfilling thing to create countless warm homes for countless people. So I gave up my dream of becoming a fashion and jewelry designer or painter. In 2005, I founded the small "XY+Z DESIGN" in Nanjing with only 20 square meters, starting my interior design journey of "creating warmth". After accumulating 5 years of design and construction experience, I stopped construction in 2010 and focused on design. In 2014, I founded "XY+Z DESIGN" in Shanghai with my husband Yaolong Wu, and gradually came to where I am today. Looking back on the experience, I couldn't help feeling emotional.

I have multiple identities in life: my mother's daughter, my daughter's mother, my husband's wife, my colleagues' colleague... I often think, if people can have so many identities, why can't space? The living room can also be a study, and the dining room is not necessarily just a place to eat. I firmly believe that the charm of design lies in its infinite possibilities. The concept of composite space design that I advocate is to break through traditional spatial boundaries, allowing each space to have multiple identities and functions, and grow naturally like an organic life. Later, perhaps I still couldn't forget my childhood dreams. I often ask myself why I can't do both interior design and fashion and jewelry design? The answer is yes. Then, in 2021, the cross-border product company "XY+Z DESIGN" named by my daughter Ruoke Wu was born. It integrates "jewelry, fashion, home, and art" into one. This year, my dream came true.

In this book, I recorded my 18-year design career and shared many of my thoughts and insights about design and life. It can be considered as an explanation for the first half of my life that is about to end—whether it is for myself, or for others, or things related to me. Therefore, you now have this book "Beyond Design" in your hands. This is my first book, please forgive me if there are any shortcomings.

A hundred solo exhibitions would be my dream in the second half of my life. I hope to present my design philosophy, my design processes, and my own understanding of beauty to more people through such solo exhibitions. Each exhibition will be a brand new beginning, a journey for me to explore beauty and life together with the exhibited and the audience.

Finally, I want to thank my parents for giving me life; thank my clients for helping me grow; thank my family and friends for their support and trust; thank my partners and colleagues for accompanying me through thick and thin; thank the editors and colleagues who supported me in preparing this book, and making it possible to meet with everyone; and thank design for making my life wonderful.

Xiwen Guo

Spring 2024, Shanghai

一

对于设计的基础理解

I. BASIC UNDERSTANDING OF DESIGN

首先，这并不是一本专业图书，更多的是关于我对设计的理解，对生活的认知。

设计于我而言，离不开理性(专业)与感性(美感)，当然能够让一个物理空间活起来，
赋予其灵魂，更是我的追求。

浅谈一下我对以下专业的基础理解，便于阐述接下来要解读的案例及理念。

First of all, this is not a professional book, but more about my understanding of design and my perception of the environment.

Design, in my opinion, cannot be separated from rationality (professionalism) and sensibility (aesthetic sense). Of course, the ability to bring a physical space to life and give it a soul is my ultimate pursuit.

Let me briefly talk about my basic understanding of the following professions, which will help me explain the cases and concepts that will be interpreted next.

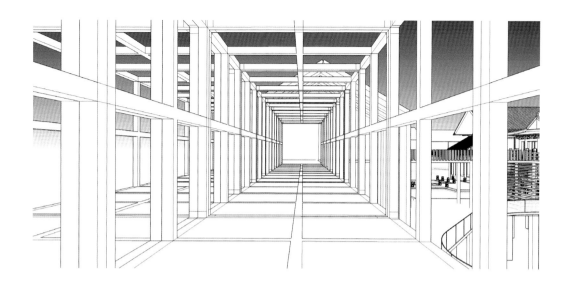

室内设计理解

MY UNDERSTANDING OF INTERIOR DESIGN

毫无疑问，设计是为了解决问题。

从专业性上来说，空间划分的合理性、制图的规范性、规范的熟知性、使用的舒适性与便利性，都应该是每一位设计人需要了解的。

设计本身是集技术、艺术于一体的，技术，我们有规可循；但艺术，就是审美了。风格可以有多种，欧式、中式、现代……于我而言，风格并不是最重要的。每个人对美的理解也不尽相同，符合定位、需求、安全、舒适、耐看，即好。

No doubt, design is for problem-solving.

From a professional perspective, the rationality of space partition, the standardization of drawing, and the familiarity of specifications, as well as the comfort and convenience of use, should be understood by every designer.

Design itself is a combination of technology and art. We have rules to follow for technology, but art is aesthetic. There are many styles: European style, Chinese style and modern style... For me, style is not the most important. Everyone has different understanding of beauty, and a good design should meet the requirements on needs, safety, comfort and durability.

建筑设计理解

MY UNDERSTANDING OF ARCHITECTURAL DESIGN

建筑承载太多，因而也相对比较沉重。

一个建筑自它落地的那一刻起，任何人都可以对其评头论足，甚至是批判。相较于室内而言，建筑需要得到更多人的认可，需要表达对于城市环境的尊重，也需要展现文化底蕴。很多建筑由于注重功能而忽略了美观，也有些建筑为了外观的标新立异而抛弃内部空间和功能的实用性，诸如此类的案例比比皆是。正因如此，外观和内部的完美合一极为必要，因此，事实上，最理想的情况是无论建筑设计还是室内设计，应该由同一家公司或同一位设计主创完成。

The burden on architecture is immense, making it relatively heavier compared to others.

From the moment a building is erected, anyone can criticize or even condemn it. Unlike interior design, architecture requires more public approval, expressing respect for the urban environment and cultural heritage. Many buildings focus on functionality and neglect aesthetics, while some sacrifice practicality and functionality of interior space for the sake of unique appearance. Numerous examples of such cases can be found. Therefore, perfect integration of exterior and interior is crucial. Hence, the most ideal situation is for a single company or person to handle both architecture and interior design.

珠宝设计理解

MY UNDERSTANDING OF JEWELRY DESIGN

自序中曾提及这是我儿时的梦想。想要设计出独特、个性的珠宝、服饰，这一点一直不曾改变。

让我觉得幸运的是，我女儿吴若可似乎也遗传了这一点，即将满十岁的她，现在的梦想也是建筑设计师、服装设计师、画家。当然，我们在相互欣赏及支持对方。我想没有谁不喜欢美，我本人也喜爱收集好看的珠宝、衣服、包包。

美是一件没有国界、没有性别的事情。怎么更美、更独一无二，是我一直想去做的事情。关键是惊艳别人的同时，先惊艳自己。这时，是否是科班出身，往往已经不重要了，你只需要好的创意，再交由专业的工厂去执行。有时我在想，或许这点和我对女儿的教育一样，如果从小就生活在各种规矩的束缚里，她是否还能保有现在的天马行空？

As mentioned in the preface, this has been my childhood dream—to design unique and personalized jewelry and clothing—which has never changed.

I feel fortunate that my daughter, Ruoke Wu, seems to have inherited this as well. She is almost ten years old now, and her current dream is also to be a fashion designer.Of course, we appreciate and support each other. I think everyone likes beauty, and I personally enjoy collecting beautiful jewelry, clothes, and bags.

Beauty is something that knows no borders or gender, and I have always wanted to create something more beautiful and unique. The key is to impress oneself before impressing others. At this point, whether one has an educational background in the field is often not crucial. One only needs good ideas and can leave the execution to professional factories. Sometimes I think that perhaps this is like the way I am raising my daughter. If she were to live her whole life under various rules and restrictions, would she still retain her limitless imagination?

产品设计理解

MY UNDERSTANDING OF PRODUCT DESIGN

产品的概念太大, 在此只能谈一谈目前我们涉猎的领域, 比如家具、家居用品等。产品存在的目的, 可以是功能, 也可以是美感, 主要取决于消费群体的需求。

当然, 好的产品可以集功能和美感于一身。所以我希望可以将美学的概念及实用的功能引入产品设计中, 有颜值又好用的产品才会被市场广泛认可。

一个建筑, 缩小, 可以是一件很优秀的产品; 一件独特的产品, 放大, 再结合落地性技术方案, 可以是一个优秀的建筑。两者之间看似有着莫大的区别, 实则充满微妙的联系。

The concept of products is too broad, so we will only discuss the fields we are currently involved in, such as furniture, home appliances, and household items. The purpose of a product can be both functional and aesthetic, depending on the needs of the consumer.

Of course, good products can combine function and beauty in one. Therefore, I hope to incorporate the concept of aesthetics and practical functions into product design, and only products that are both visually attractive and functional will be widely recognized by the market.

A building, when scaled down, can be an excellent product; a unique product, when scaled up and combined with on-site technical solutions, can become an excellent building. While the two may seem vastly different, they are actually full of subtle connections.

二

实践与方法

II. PRACTICE AND METHODS

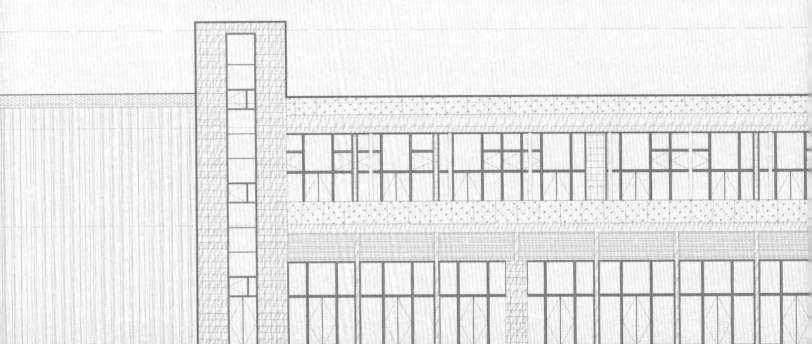

房地产示范区复合人文功能空间

COMPOUND HUMANISTIC FUNCTION SPACE OF MARKETING CENTER

书店系列

BOOKSTORE SERIES

从事房地产示范区销售中心设计很多年,经历过多次几个月从一片荒地到项目呈现,再到楼盘售罄,临时销售中心拆除。每每这个时候, 总是心疼无比。也许是心疼甲方投入的费用, 也许是心疼我的设计。这个时候就想, 怎样才能完成销售使命后, 还能让她"活"久一点。

2015 年, 第一个"不纸书店"诞生于南京。从功能复合化上的"不止", 到室内出现太多纸的"不纸", 再到 2016 年成都的"不纸书店 2.0"。她们掀起了地产界销售中心复合功能的热潮, 至今。

不纸书店系列算是我的代表作之一。成功延长了她们的寿命 (不纸书店南京, 成功转型为永久建筑, 并且以书店形式一直存在), 这让我万分欣慰。当然, 长寿是一方面, 而对于空间的设计个性, 同样重要。"健康、绿色、时尚、人文", 是我一直秉持的设计理念。所以, 以"纸"为主要材料, 与店中的"书纸"共舞。这如同一个人, 美丽皮囊和有趣的灵魂, 都缺一不可。与此同时, 她们都有相同的特点: 复合、多元、人文。

I have been designing sales centers for real estate demonstration areas for many years. I have experienced several months' journey from a barren land to project presentation, and then to the demolition of the temporary sales center after the property is sold out. Every time, I feel extremely distressed. Maybe it's because I feel sorry for the cost invested by the client, or maybe it's because I feel sorry for my design. At this time, I always think about how to make it "live" longer after completing its sales mission.

In 2015, the first BUZHI BOOKSTORE was born in Nanjing. From the functional diversification of "BUZHI BOOK-STORE" to the appearance of too much paper in the interior design, to the "BUZHI BOOKSTORE 2.0" in Chengdu in 2016, they have ignited a trend of compound functions in real estate sales centers.

The BUZHI BOOKSTORE series is one of my representative works. In addition to successfully prolonging their life (BUZHI BOOK-STORE in Nanjing has been transformed into a permanent building and has been operating as a bookstore ever since), which makes me extremely pleased. "Healthy, green, fashionable, and humanistic" is the design philosophy that I have always adhered to. Therefore, I use "paper" as the main material and dance with the "book paper" in the store. This is like a person, both a beautiful appearance and an interesting soul are indispensable. At the same time, they have the same characteristics: complex and diverse.

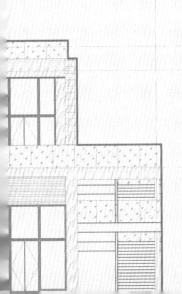

不纸书店升级书院系列
ACADEMY SERIES

今天的书店是让人放松、愉悦的地方，而古时的书院具有求学、授课、展览功能，是庄严肃穆的。学子们在此学习、塑造及追求自己未来的人生。我认为，任何风格都应建立在定位之上。而对于项目定位高端、户型产品面积段大，且本身品牌定位高的产品，书店是无法承载的。基于这样的考虑，我对书店的定位进行了升级，用书院、书画院等复合庄严的空间来定义该类设计的空间属性。中间还有一些小插曲，让我非常赞同"人的欲望是无止境的"。对设计，我想我也是这样无止境。目前国内的大部分项目，都是工期赶，预算紧。那么售楼部的寿命延长的同时，能不能再快一点？或者内部装修的材料可以支持拆装，二次使用？因为越来越多的售楼部已经不再是临建，而是用配套用房进行销售空间展示。此时的建筑业，已经被"装配式"覆盖。无论工期还是预算，都因此得到了很好的控制及改善，但"室内"此时对"装配"应用得还不太广泛。

在我先生的影响下，借由万科无锡天一玖著项目，我们也在思考如何推动室内的装配式变革。但实际上发现，每一个新的技术或想法，在最初推动的时候往往困难重重。最终，该项目还是没有达到预期的装配率，但还是让我们在室内装配这条路上迈出了一小步。后面关于室内装配这个话题，我们再来细说。

Today's bookstores are places where people can relax and enjoy themselves, while old-style academies, with their functions of study, teaching, and exhibitions, were solemn and dignified. Students learned, shaped themselves, and pursued their future at these academies. I believe that any style should be built upon a strong positioning. For projects that are positioned as high-end, with large-sized products and high brand positioning, bookstores cannot meet this demand. With this in mind, I upgraded the positioning of bookstores by defining the spatial attributes of such designs as compounding dignified spaces such as academies and painting academies. There were some small anecdotes in between that made me fully agree with the idea that "human desire is infinite." I believe that for design, the same holds true. Most of China's current projects are characterized by short construction periods and tight budgets. As such, the lifespan of a sales center can be extended, while the interior decoration materials can support disassembly and reuse. This is because more and more sales centers are no longer temporary structures but are sales spaces that are built with supporting facilities. The construction industry is already covered by "modular construction", which has effectively controlled and improved the construction period and budget. However, the application of "modular construction" in interior is not very extensive at this time.

Under the influence of my husband, with the help of Vanke Tianyi Jiuzhu project, we are also thinking about how to promote the reform. But in fact, it is found that every new technology or idea is often difficult to be accepted at the beginning. In the end, the project did not meet the expected assembly rate, but it still allowed us to take a small step on the road of interior modular assembly. Let's talk about the topic of interior modular in detail later.

联合性艺术空间系列
GALLERY SERIES

当然并不是所有的定位、客群都适合以"书"诠释。艺术馆给人更丰富、年轻的感受，也是非常不错的选择。而销售中心本身就是展示空间，和艺术馆的属性定义不谋而合。我的爱好很广泛，旅行、看书、看展、音乐、咖啡、茶、酒……长期以来，"看展"已经成为一种习惯。将销售展示空间与艺术展示空间融合，对售楼部功能复合化的拓展，也带来了非常多的延展作用。同时，也为设计美术馆、博物馆类独立艺术空间奠定了基础。

Of course, not all positioning and customers ware suitable for "book" interpretation. The art museum gives people a richer and younger feeling and is also a very good choice. The sales center itself is a display space, which coincides with the attribute definition of the art museum. My love is very extensive. Travel, reading, exhibition, music, coffee, tea, wine... "Exhibition watching" has become a habit for a long time. The integration of sales exhibition space and art exhibition space has also brought a lot of extension effects to the expansion of the sales department's functional integration. At the same time, it also laid a foundation for designing independent art spaces such as art galleries and museums.

不纸书店 1.0
BUZHI BOOKSTORE 1.0

2015 "永隆·星空间杯" 江苏省室内设计、陈设设计大奖赛铜奖

■

业主单位：朗诗地产 - 朗诗南京熙华府销售中心
项目地点：江苏南京
面积：700 平方米
完工时间：2016
主要材料：白蜡木、瓦楞纸、艺术漆、免漆板

Client: Landsea Property-Landsea Xihua Community Sales Center, Nanjing
Project Location: Nanjing, Jiangsu Province
Area: 700 sqm
Completion Time: 2016
Materials: Ash wood, corrugated paper, art lacquer, lacquer-free panels

这里是一个书店：纸订成书，书摞成馆；数万册的藏书，纸花瓶、纸凳、纸桌、纸灯、纸桶……处处均为纸的元素；多到极致，即是无；纸，即不纸。

这里又不只是一个书店：它是一个融合了茶、咖啡、轻餐、艺术品鉴、骑行体验、生态农场等多功能的复合空间。

不纸书店，不只是书店。

This is a bookstore: paper stapled into books, books stacked into a library; tens of thousands of books, paper vases, paper stools, paper tables, paper lamps, paper buckets... Everywhere there are elements of paper; so much that it is nothing, and paper that is not paper.

It is not just a bookstore: it is a multifunctional space that combines tea, coffee, light meals, art appreciation, cycling experience and eco-farm.

BUZHI BOOKSTORE is more than just a bookshop.

奔波、忙碌……每天快节奏的生活挤压着我们。在这样的状态下，作为这次书店的设计者，我思考，在每个人的内心深处究竟需要一个什么样的空间，让我们可以暂时逃离这样的状态；而置身在这 700 平方米的玻璃房中，我们如何才能借着清晨或黄昏的阳光，思考自己能为这个浮躁的社会做些什么。

安静、温暖、质朴、舒适、小憩、聊天、咖啡、茶、阅读…… 是的，她该是这样的性格。

The fast-paced daily life of running around and being busy squeezes us. In such a state, as the designer of this bookstore, I pondered what kind of space is needed deep inside everyone's heart so that we can escape from such a state for a while; and how, being in this seven hundred square meters glass room, we can take advantage of the early morning or dusk sunlight to think about what we can do for this impetuous society.

Quiet, warm, rustic, comfortable, napping, chatting, coffee, tea, reading... Yes, that's how she should be characterized.

她是艺术的、震撼的：巨幅米开朗琪罗的《创世纪》，置于天顶之上，当上帝向亚当伸出手的那一瞬间，便开启了你的心阅之旅；意大利 Abstruco 艺术地坪，如斯美塔那的交响乐"沃尔塔瓦河"般，在地面激昂、奔腾地流淌。

她是精致的、人文的：鹦鹉螺的自然、完美、精准的比例，在这里，在灯上，在桌前，在门间……随处可见；健康、骑行、美食、人文，她在诉说一种情怀。

She is artistic and breathtaking: the huge Michelangelo's "*Creation of Adam*", placed on the ceiling, opens your heart journey at the moment when God reaches out to Adam; the Italian Abstruco art floor flows on the ground as passionately as Smetana's symphony "*The Moldau*".

She is delicate and humanistic: the natural, perfect and precise proportions of the nautilus can be seen everywhere here, on the lamps, tables, and doors; healthy, cycling, gastronomy, and humanistic, she speaks of a kind of sentiment.

立面图 Elevation

1F 平面图 1F Plan

0 1m 2m   4m   10m

1 门厅 Foyer
2 水吧区 Water Bar Area
3 餐区 / 阅读区 Dining / Reading Area
4 阅读区 Reading Area
5 儿童阅读区 Children's Reading Area
6 卫生间 Toilet
7 外摆区 Swing Area
8 生态农场 Eco Farm
9 健康体验区 Wellness Experience Area

2F 平面图（阅读模式）2F Plan (Reading Mode)

1 阅读区 1 Reaning Area 1
2 室外阅读区 Outdoor Reading Area
3 艺术品鉴区 Art Appreciation Area
4 前言墙 Foreword Wall
5 自助咖啡区 Self-Service Coffee Area
6 门厅上空 Void above Foyer

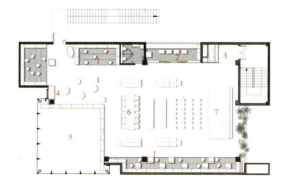

2F 平面图（沙龙模式）2F Plan (Salon Mode)

1 阅读区 1 Reaning Area 1
2 室外阅读区 Outdoor Reading Area
3 前言墙 Foreword Wall
4 自助咖啡区 Self-Service Coffee Area
5 门厅上空 Void above Foyer
6 授课区 Teaching Area
7 沙龙区 Salon Area

房地产示范区复合人文功能空间 COMPOUND HUMANISTIC FUNCTION SPACE OF MARKETING CENTER

她是绿色的、生态的：垂直绿化、立体循环，净化着我们快要成焦炭的身体，试图拼命创造一屋干净的空气，让我们可以用力地呼吸；纸元素、原木的"零甲醛"概念运用，共同打造环保、绿色的生态空间。

她是经典的、时尚的：阿切勒·卡斯蒂格利奥尼（Achille Castiglioni）为PLOS（"美国科学公共图书馆"）设计的灯、菲利普·帕特里克·斯塔克（Philippe Patrick Starck）为KARTELL设计的椅子等国际知名大师的家具、饰品，为这个空间注入了经典的血液；色彩的运用更体现着她的年轻、时尚与活力。

She is green and ecological: vertical greening and three-dimensional circulation purify our bodies that are about to turn into charred coal, trying to create a clean air in the house so that we can breathe deeply; the concept of "zero formaldehyde" is applied to paper elements and raw wood to create an environmentally friendly and green ecological space.

She is classic and fashionable: the internationally renowned masters' furniture and decorations such as Achille Castiglioni's light design for PLOS and Philippe Patrick Starck's chair design for KARTELL inject classic blood into this space; the use of colors also reflects her youthfulness, fashion, and vitality.

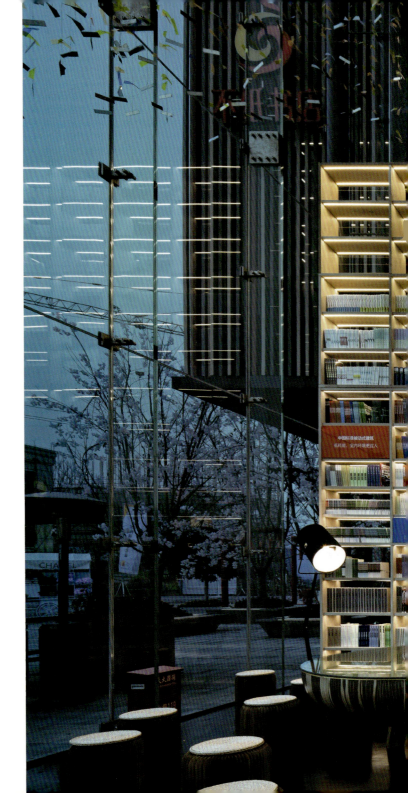

她是质朴的、慵懒的：山里的老农手工编织的蒲草垫，仿佛还能触摸到岁月的痕迹，不带修饰、还原真实；落地灯下的懒人沙发，一杯咖啡或清茶，是我们懒懒放松的地方。

我们提倡读书，读好书，读纸的书，回归一种健康的生活方式。

不纸书店，不只书店。

She is rustic and lazy: the old farmers in the mountains hand-weaved the futon cushions, as if you can still touch the traces of the years, without embellishment; the lazy sofa under the floor lamp, a cup of coffee or tea... it is a cozy place for us to relax.

We promote reading, reading good books, reading paper books, returning to a healthy lifestyle.

BUZHI BOOKSTORE, not just a bookstore.

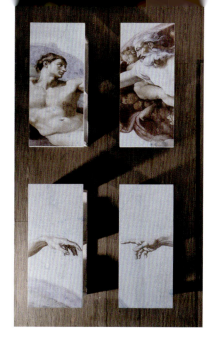

房地产示范区复合人文功能空间 COMPOUND HUMANISTIC FUNCTION SPACE OF MARKETING CENTER   023

不纸书店 2.0
BUZHI BOOKSTORE 2.0

2016 金堂奖年度最佳零售空间设计奖
2016 地产设计大奖中国组委会金奖
2016 "永隆·星空间杯" 江苏省室内设计大赛银奖
2017 第二十届中国室内设计大奖赛零售商业工程类入选奖
2017 入选中国室内设计年鉴

■

业主单位：成都太行瑞宏房地产开发有限公司·朗诗集团 - 太行瑞宏·朗诗金沙城售楼部
项目地点：四川成都百仁公园
面积：2383 平方米
完工时间：2017

Client: Chengdu Taihang Ruihong Real Estate Development Co., Ltd.
(Landsea Group) /Landsea Jinsha Town Sales Center, Chengdu
Project Location: Bairen Park, Chengdu, Sichuan Province
Area: 2,383 sqm
Completion Time: 2017

继南京朗诗不纸书店之后，升级版不纸书店 2.0 正式上线。在传统售楼处基础之上，汲取传统书店的优势，设计融入 "纸" 元素，构造 "售楼 + 办公 + 书店" 复合体验的方式，打造集绿色、健康、科技、人文、时尚五大主题于一体的文化聚集地。

Following the Buzhi Bookstore in nanjing, the upgraded version of Buzhi Bookstore 2.0 is officially launched. Based on the traditional sales office, it draws on the advantages of traditional bookstores, and is designed to incorporate the "paper" element and offer a different experience of "sales building + office + bookstore" to create a cultural gathering place with five themes: green, health, technology, humanity and fashion.

走进不纸，首先映入眼帘的是时光隧道，仿佛是从物质世界穿越进精神空间的一个通道，为人们指明走向人文的路，让人们重归丰沃的精神土壤，给即将搁浅的灵魂更多生机。不论圣洁高贵的白色亮光还是生机勃勃的植物墙，无一不是视觉盛宴。

Upon entering the bookstore, the first thing that catches the eye is the "Time Tunnel" as if it were a passage from the material world to the spiritual realm, pointing the way to the humanities and leading people back to a fertile spiritual soil, providing more vitality to souls that are about to be stranded. Whether it is the sacred and noble white light or the vibrant plant wall, every detail is a feast for the eyes.

1F 平面图 1F Plan

0 2.5m 5m   10m   25m

1 销售区 Sales Area
2 模型台 Model Table
3 销售大厅 Sales Hall
4 洽谈区 Negotiation Area
5 体验区 Experience Area
6 走廊 Corridor
7 书店 Bookstore
8 服务台\接待 Reception

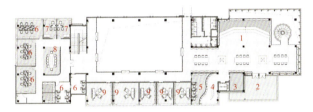

2F 平面图 2F Plan

1 书店 Bookstore
2 露台 Terrace
3 书库 Library
4 儿童区 Children Area
5 财务办公 Finance Office
6 办公室 Office
7 经理室 Manager's Office
8 会议室 Meeting Room
9 独立阅读室 Reading Room

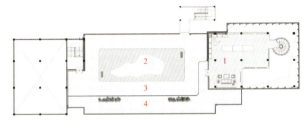

3F 平面图 3F Plan

1 书法厅 Galligraphy Hall
2 迷你高尔夫练习场 Mini Golf Club
3 卡丁车赛道 Go-Kart Track
4 无人机体验区 / 烧烤区 / 屋顶酒会 Drone Experience Area/
  Barbecue Area/ Rooftop Cocktail Party

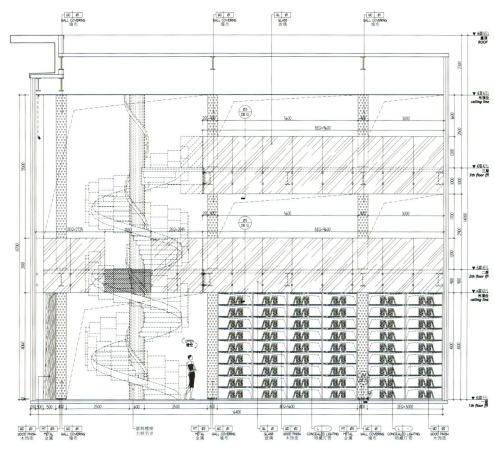

立面图 Elevation

房地产示范区复合人文功能空间 COMPOUND HUMANISTIC FUNCTION SPACE OF MARKETING CENTER **033**

川流不息，金沙不纸；生命不息，学习不止。不纸书店所能带给我们的更是对于生命的感悟。

"曾经跨过山河大海，也穿过人山人海"，来金沙城，见识与众不同的星河璀璨。

不纸书店，不只是书店。你可以在其中享受阅读、发呆、聊天、小憩、咖啡与安逸。一个蒲团、一杯咖啡、一本好书，就是一个美妙下午。这里不同于任何一个传统售楼处。

The flow of water never stops, and neither does the golden sand, and life is a continual process of learning. Buzhi bookstore brings us a deeper understanding of life.

Having traveled across mountains and seas and through crowds of people to arrive in jinsha (golden sand) town, we witness in Buzhi Bookstore a brilliant galaxy that sets itself apart.

Buzhi Bookstore is more than just a bookstore; it offers a place for reading, daydreaming, chatting, relaxing, having coffee, and feeling at home. With a comfortable cushion, a cup of coffee, and a good book, an afternoon becomes splendid. This place is unlike any traditional sales office.

撒哈拉沙漠 Sahara Desert
代表区域：非洲 Typical region: Africa
代表动物：骆驼 Typical animal: camel
代表气候：风 Typical weather: wind

阿尔卑斯山 Alps
代表区域：欧洲 Typical region: Europe
代表动物：高地山羊 Typical animal: Highland goat
代表气候：雪 Typical weather: snow

亚马孙雨林 Amazon Rain Forest
代表区域：美洲 Typical region: America
代表动物：鳄鱼 Typical animal: crocodile
代表气候：雾 Typical weather: fog

太平洋 Pacific Ocean
代表区域：大洋洲 Typical region: Oceania
代表动物：鲸鱼 Typical animal: cetacean
代表气候：雨 Typical weather: rain

蒙古大草原 Mongolian Steppe
代表区域：亚洲 Typical region: Asia
代表动物：草原骏马 Typical animal: steed
代表气候：晴 Typical weather: shine

房地产示范区复合人文功能空间 COMPOUND HUMANISTIC FUNCTION SPACE OF MARKETING CENTER

万科天一玖著书院

VANKE TIANYI JIUZHU ACADEMY

2018 "永隆·星空杯" 江苏省室内设计大赛银奖

2018 金堂奖售楼处 / 样板间年度优秀

业主单位：无锡市辰万房地产有限公司 - 万科无锡天一玖著销售中心
项目地点：江苏省无锡市惠山区
面积：955 平方米
完工时间：2017

Client: Wuxi Chenwan Real Estate Co. /Vanke Tianyi Jiuzhu Sales Center, Wuxi
Project Location: Huishan District, Wuxi, Jiangsu Province
Area: 955 sqm
Completion Time: 2017

万科天一新城位于无锡惠山区的关键地段，同时辐射天一、西漳、刘潭三个关键片区，周边学区环绕，交通便利。如何与周边建筑和文化生活融为一体，这是我们考虑的关键要素之一。

在无锡这片土地上，曾经诞生过很多文人名士，也创立了许多知名书院。不愿去打破这浓厚的文化氛围，于是，我们选择留下，书院的概念就此悄然而生。

Vanke Tianyi New Town is located in a key location in the Huishan District of Wuxi, adjacent to Tianyi, Xizhang and Liutan (three key districts), surrounded by school districts and convenient transportation. How to integrate with the surrounding architecture and cultural life is one of the key elements we considered.

The land of Wuxi has been home to many literary figures and famous scholars, and many famous academies have been founded. Not wanting to break this rich cultural atmosphere, we chose to retain it and the concept of the academy was born.

一門美學 融粹百家之生活哲理

一個時代 延續文脈之至善至譽

一份自然 宛如初見之天地美好

一種卓越 洞悉城市之強者雄心

有氣有化 宜室宜家 金鑲玉勒 桃花灼灼

投我以木李 報之以瓊玖

此玖 永以為好

文化是一个城市的中流砥柱，而我们想让天一成为这座城市的文化沟通桥梁。教学，讲究"传道、授业、解惑"，而当销售中心融入书院气质，一切又变得与众不同。在这里，不仅提供展览和藏书，还会建立国学讲堂，与周边学区做互动交流，让孩子们从小接受国学素养的熏陶。为了让更多人触摸到艺术品以及传统工艺品，这里还会搭建无锡艺术家的交流平台，开放销售渠道，让文化和技艺不止存在于历史的长河中。

延续东方文化的设计理念，前厅以中庭景观为景，以文学名画为墙，整个空间并没有太多的遮挡，通透明亮的设计风格倒与中国传统园林景观有异曲同工之妙。

走廊内，"官帽灯"照亮下方的拴马柱。"官帽灯"指引着锦绣前程，拴马柱则警示世人戒骄戒躁，因为只有怀着一颗纯粹的求学之心才能真正领悟文化的真谛。墙上使用超白低反的玻璃材质，以古文的方式缓缓讲述着万科的文化和历程，好似回归几千年前，听着先生讲解着书中的故事。

Culture is the backbone of a city, and we want Tianyi to be a cultural bridge for the city. Teaching is all about "preach, teach and solve puzzles", but when the sales center is integrated with the atmosphere of a school, everything becomes different. Not only does it offer an exhibition and book collection, but it also establishes a national education lecture hall and interacts with neighboring school districts, so that children can be educated in national education from an early age. In order to make art and traditional crafts more accessible to more people, we will also build an exchange platform for local artists and open up sales channels, so that culture and skills do not just exist in history.

Continuing the design concept of oriental culture, the vestibule is decorated with a landscaped atrium and a wall of famous literary paintings, the whole space is not overshadowed and the bright design style is similar to that of traditional Chinese gardens.

In the corridor, the "official hat lamp" illuminates the stanchion underneath. The "official hat lamp" guides you to a bright future, while the stanchion warns you to abstain from arrogance, because only with a pure heart of learning can you understand the true meaning of culture. The wall is made of ultra-white, low-reflective glass, and slowly tells the culture and history of Vanke in an ancient style, as if you are returning to thousands of years ago, listening to the mentor explain the story in a book.

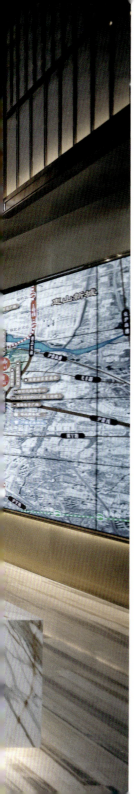

这里可以是阅读空间，也可作为文学讲堂。就像项目 logo 一样，多重书籍堆叠，从王朝历史到近代文化，愿你能静静享受这种东方古典之美。

天一·玖著，我们赋予其书卷气息以及古风韵味，让销售中心不止于商业。

如今，落成后的它，愿带给你无尽享受……

It can be a reading space or a lecture hall. Just like the logo of project, multiple books are stacked, from dynastic history to modern culture, and may you quietly enjoy this classical oriental beauty.

Here we created a bookish atmosphere as well as an ancient charm to make the sales center more than just a business.

Now, after its completion, we wish to bring you endless enjoyment...

1F 平面图 1F Plan

0 3m 6m   12m   30m

1 主入口 Main Entrance
2 礼宾 Concierge
3 大堂 Lobby
4 文化展示 Cultural Exhibition
5 销控台 Sales Control Station
6 户型沙盘 Apartment Type Model
7 模型台 Model Table
8 独立阅读室 Independent Reading Room
9 洽谈区 Negotiation Area
10 水吧 Water Bar
11 贵宾室 VIP Room
12 儿童游乐室 Dollhouse
13 洗手间 Toilet

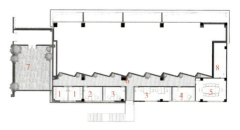

2F 平面图 2F Plan

1 更衣室 Dressing Room
2 物业办公室 Property Office
3 办公室 Office
4 经理室 Manager's Office
5 会议室 Meeting Room
6 走廊 Corridor
7 屋顶花园 Roof Garden
8 储藏室 Storage Room

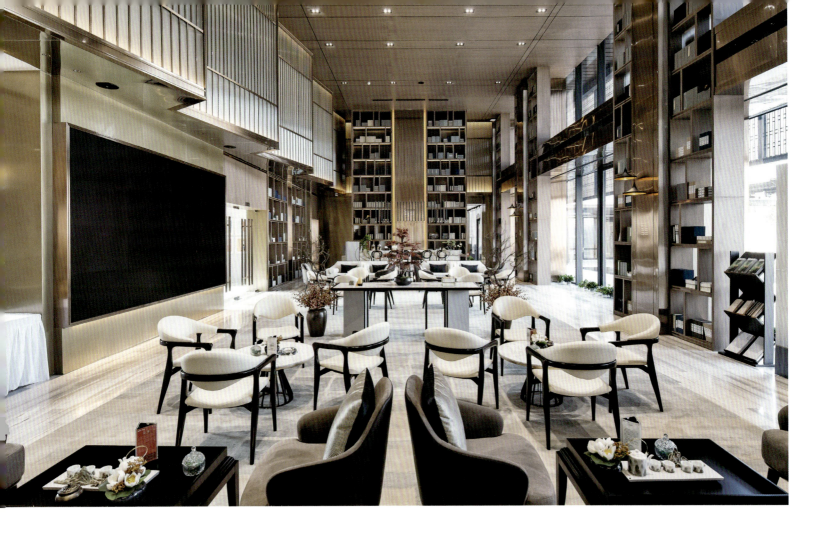

销售区域，使用木饰面打造全新的墙壁形态，好像有无数本书整齐排列于墙面，折线形的设计又增添了整体风格的灵动性，巧妙地将销售中心从传统呆板的模式中解救出来。而周身镜面不锈钢的多重使用，以其反射效果拉高了整个空间感。

所有的沙发都使用了阿玛尼的面料，极大提升了整个空间的品质感。这里不仅为顾客提供舒适的交流空间，也可以是高雅别致的休息场所。

In the sales area, the use of wood veneer creates a new wall form, as if there are countless books neatly arranged on the wall, and the bending line design adds to the overall dynamism of the style, cleverly rescuing the sales center from the traditional stagnant model. The multiple use of mirrored stainless steel around the perimeter heightens the sense of space with its reflective effect.

The use of Armani fabrics for all the sofas adds a sense of quality to the whole space. What we wanted to give our customers was a comfortable space for communication, but of course it could also be an elegant place to rest.

销售模式 Sales Model

授课模式 Teaching Mode

在"设计理解"中有提到，天一玖著项目是我对"室内装配式"的探索。实际建成的项目并非与最初的构思相符。因工期限制，项目没有时间去试错及造新，这在当时让我十分遗憾。此时无比庆幸能出这本书，让原构思得以呈现，哪怕是以出现在书中的形式。虽然最后并没有实现这个方案，但也为之后我的"室内装配式"打下了很好的基础。

展览模式 Exhibition Mode

As mentioned in "Design Understanding", the Tianyi Jiuzhu project is my exploration of indoor assembly. The actual project that was built was not originally conceived. Due to the time limit, the project did not have time for trial and error and new creation, which made me very sorry at that time. At this time, I am very glad to be able to produce this book, so that the original idea can be presented, even in the form of a book. Although this scheme was not realized in the end, it laid a good foundation for my "indoor assembly" in the future.

最初开始这个项目时，就在脑海中不断告诉自己，这个项目已经不再是书店能承载的了。一是因为项目产品定位高，同时面积段大，客群不同。二是因为"万科"。这在古代就是普通文人和士大夫的区别了。我们从万科的发展史：从最初的探索，到后面的风格独定，居于高位，业界无出其右；对应书院求学的递进关系：从初入书院时的探索求知，到能够执笔道章，最后悟至大道至简。以此，将该项目定位为书院。而书院，在古代则具有展览、藏书、讲学的功能。所以，最初的方案我们以这些功能展开，同时，融入当代必不可少的"智能"。通过书柜的隐藏式轨道和不同场景设定的程序进行电动位移、家具的不同堆叠方式、电动幕布的调整、灯光及投影的配合，实现售楼、讲堂、展览、放映、酒会等完全不同的功能场景。

When I first started this project, I kept telling myself that it was no longer fit for the bookstore. First, because the project positioning is high, while the area is large, the customer base is different. The second is because of "Vanke". This was the difference between ordinary literati and scholar-officials in ancient times. We started from the development history of Vanke: from the initial exploration, to the unique style. This is related to the experience of studying in an academy: from the exploration of knowledge in the beginning, to being able to write, and finally to be a master. In this way, the project is positioned as an academy. Chinese academies, in ancient times, had the functions of exhibition, collection of books and lectures. Therefore, the initial solution we started with these functions. At the same time, we tried to incorporate the essential contemporary "intelligence". With the hidden tracks in the bookcases and the set programs, we can realize electric displacement, different stacking ways of furniture, adjustment of electric curtain, lighting and projection coordination. In this way, sales center, lecture halls, exhibitions, screenings, wine parties and other completely different functional scenes are combined.

光影展览模式 Light Exhibition Mode

酒会模式 Reception Mode

朗诗金堂书院
LANDSEA JINTANG ACADEMY

2018 "永隆·星空杯" 江苏省室内设计大赛优秀奖
2018 中国室内设计大奖赛入选奖

■

业主单位：朗诗集团 - 朗诗成都金堂销售中心
建筑设计：吴耀隆
景观设计：重庆创合园林设计有限公司
项目地点：成都市金堂县
面积：950 平方米
完工时间：2018

Client: Landsea Group/Landsea Jintang Sales Center, Chengdu
Architectural Design: Yaolong Wu
Landscape Design: Chongqing Chuanghe Landscape Design Co.
Project Location: Jintang County, Chengdu, Sichuan Province
Area: 950 sqm
Completion Time: 2018

金堂书院是 XY+Z DESIGN 与朗诗合作的又一力作。我们因不纸书店结缘，此次也延续复合型空间的设计理念，集成都的人文以及悠长的文化为一体，将书店升华成书院。

Jintang Academy is another masterpiece of cooperation between XY+Z DESIGN and Landsea. We have developed a connection with Buzhi Bookstore, and this time we are continuing the design concept of a mixed-use space, integrating the humanities and the long-standing culture of the city and turning the bookstore into an academy.

玻璃盒子的入口处，运用垂吊着的点状构件，形成完整的"沉思者"。背景选用发光阳光板，将这个空间艺术打造成一个具有文化底蕴的视觉亮点。几何元素的拼接，通透的视觉效果与水面、植物的搭配。动与静、繁与简，共同演绎富有节奏的建筑美感。

The entrance to the glass box is completed by the use of dangling dotted elements to form the "contemplator". Luminous sunlight panels are used in the background, transforming the space into a visual highlight with a cultural heritage. The combination of geometric elements, transparent visual effects and the matching of water and plants... Movement and stillness, complexity and simplicity, together interpret a rhythmic architectural aesthetic.

水吧台发光夹丝玻璃的运用，为整个古色古香的氛围增添了一份现代艺术气息。签约区则以木色为主，选用与水吧背景相同的透植玻璃，用现代简洁的手法将绿色生态融入室内每个细节。

阅读室是书香气最为浓厚的地方。中式元素随处可见，水墨山水、木制家具……顶部与竹简型壁纸相匹配的金属条延伸至书架，安谧、舒适的氛围，让人逐步抽离现实，乐享书中自在。

寻一处隅角，阅一部古籍，吟一篇诗词，做一回文人。

诗词歌赋，低唱浅斟，愿还你一方闹市中的静谧。

The use of luminous laminated glass in the water bar adds a modern artistic touch to the ancient atmosphere. The signing area is dominated by wood, with the same translucent glass as the backdrop of the water bar, using a modern and simple approach to integrate green ecology into every detail of the interior.

The reading room is the place where the aroma of books is strongest. Chinese elements can be found everywhere, from landscape paintings to wooden furniture... The metal strip on the top matching the bamboo wallpaper extends to the bookshelves, providing a tranquil, comfortable atmosphere where one can gradually withdraw from reality and enjoy the freedom of books.

Find a corner, read an ancient book, recite a poem, and be a literary person.

We hope to bring you to the quietness of the city.

洽谈区沿用极具现代感的表现手法，还原传统书院独具的古典气息。即使烈日高照，室内也有效阻隔了阳光的直射，归功于侧边吊顶的拉膜结构。

这是一处创新，以拉膜模拟书院帘布的质感，柔化太阳光线，营造室内柔和的文化氛围。

The negotiation area adopts a highly modern style while restoring the classical atmosphere unique to traditional academies. Even under the scorching sun, the interior effectively blocks the direct sunlight thanks to the stretched membrane structure of the side ceiling.

This is an innovative design that simulates the texture of the academy curtain with stretched membranes, softening the sunlight and creating a gentle literary atmosphere indoors.

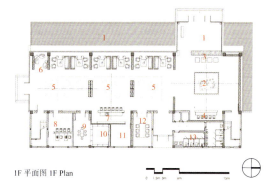

1 接待区 Reception Area
2 沙盘 Sand Table
3 户型沙盘 Model Sand Table
4 销控台 Sales Control Station
5 洽谈 / 书吧 Negotiation/Book Bar
6 阅读角 Reading Area
7 水吧 Water Bar
8 签约房 Signing Room
9 儿童娱乐区 Children's Area
10 泡泡池 Bubble Pool
11 储物间 Locker Room
12 独立阅读室 Reading Room
13 洗手间 Toilet

1F 平面图 1F Plan

1 项目总经理办公室 General Manager's Office
2 营销办 Sales Office
3 储藏室 Storage Room
4 会议室 Meeting Room
5 资料室 Reference Room
6 监控室 Monitoring Room
7 物业办 Property Office
8 物业休息室 Property Lounge
13 洗手间 Toilet

2F 平面图 2F Plan

上河艺术馆
SHANGHE ART GALLERY

2020 A'Design Award 设计大奖银奖
2021 "永隆·星空间杯"江苏省室内设计大赛银奖
2020 金堂奖年度杰出售楼处

■

业主单位: 朗诗·西安-朗诗开封汴京西华府销售中心
景观设计: 上海易境环境艺术设计有限公司
建筑设计: 上海柏涛建筑设计咨询有限公司
项目地点: 河南开封
面积: 1050 平方米
完工时间: 2020

Client: Landsea Xi'an/Landsea Xihua Community Sales Center, Kaifeng
Landscape Design: EGS Design
Architectural Design: Shanghai PTArchitects
Project Location: Kaifeng, Henan Province
Area: 1,050 sqm
Completion Time: 2020

上河艺术馆的设计将当地的历史人文与"人本、阳光、绿色、科技"的理念有机地融合为一体。昔日汴京城的建筑、家具、装饰、书画、民间技艺等元素在空间里重新交汇,全新演绎一场跨越千年的东京梦华。

The design of the Shanghe Art gallery organically integrates local history and humanity with the concepts of humanism, sunlight, greenery and technology. The architecture, furniture, decoration, painting and calligraphy, folk skills and other elements of the former Bianjing city are reunited in the space, and a new interpretation of the ancient city spanning a thousand years is created.

开封，古称汴京，素有"八朝古都"之称，孕育了上承汉唐、下启明清、影响深远的"宋文化"。大宋王朝是开封历史上和中国封建史上最为辉煌的时代。史书更以"八荒争辏，万国咸通"来描述当时大宋都城开封的繁华。而今家喻户晓、妇孺皆知的遗世名作《清明上河图》，描绘的便是清明时节北宋东京开封城汴河两岸的繁华、热闹的景象和优美的自然风光。本案以《清明上河图》作为开封历史文化的象征，结合当地巧夺天工的剪纸技艺，运用现代金属雕刻技术及中国红将其全新演绎，成为整个空间里最大的亮点。

Kaifeng, known as Bianjing in ancient times, is known as the ancient capital of the eight dynasties and has nurtured the far-reaching Song culture, which was inherited from the Han, Tang, Ming and Qing dynasties. The Song Dynasty was the most glorious era in Kaifeng's history and in the history of feudalism in China. The Qingming Shanghe Tu, a masterpiece that is now a household name and known to all women and children, depicts the prosperous and lively scene and beautiful natural scenery on both sides of the Bian River in Kaifeng, the capital city of the Northern Song Dynasty during the Qingming Festival. The case uses the Qingming Shanghe Tu as a symbol of Kaifeng's history and culture, combined with the skillful local paper-cutting techniques, and uses modern metal engraving techniques and Chinese red to reinterpret it, making it the biggest highlight of the whole space.

空间布局上，首层规划主要满足售楼处的商业属性，从入口、接待、沙盘展示，到前洽、深洽、销控等区域销售，动线一气呵成。

二层艺术长廊则更侧重空间的艺术属性。顶部设计顺应建筑原有的人字形结构，顺势而为的处理手法中暗藏天人合一的东方智慧。功能上除了日常艺术品展示之外，也可用于举办特展。同层的多功能厅既可作为会议室，亦可举办国学讲堂。

我们通过建筑立面的大面积落地玻璃，运用借景的手法将庭院内的曲径流觞尽收眼底。以水为镜，建筑自身的倒影亦成为景的一部分，所谓你中有我，我中有你，内外交融，对立而又统一，传达了古老而深邃的东方哲学。

售楼处大厅地面整体由深色保加利亚灰石材铺设而成，使通透的大厅显得现代又不失大气。同时与二层空间形成明暗对比，利用视觉感受形成上古与当今的对立统一。

In terms of space layout, the first floor is planned to meet the commercial attributes of the sales office, from the entrance, reception, sand table display, to the first contact, deep contact, sales control and other sales lines in one go.

On the sencond floor, the art gallery focuses on the artistic attributes of the space. The design of the roof follows the original herringbone structure of the building and conceals the oriental wisdom of the unity of nature and man in the way it is handled. In addition to the daily display of artworks, it can also be used for special exhibitions. The multi-functional hall on the same floor can be used as a meeting room or as a lecture hall for Chinese studies.

Through the use of large floor-to-ceiling glass on the façade, the designers have used a borrowed view of the landscaping in the courtyard. With the water as the mirror, the reflection of the building itself becomes part of the scene. the inside and outside are intertwined, opposing but unified, and this also conveys the ancient and profound oriental philosophy.

The floor of the sales office is made up of dark Bulgarian gray stone, giving the airy hall a modern appearance. It also contrasts with the second floor space, creating a visual sense of unity between the ancient and the modern.

效果图 Rendering

1 售楼部入口 Entrance
2 水景 Water Feature
3 区位地图 Location Map
4 大型沙盘 Large Sand Table
5 销控台 Sales Control Station
6 前台招待 Recetionist
7 洽谈区 Negotiation Area
8 水吧 Water Bar
9 收银 Cashier
10 财务办公室 Finance Office
11 儿童娱乐区 Chirdren's Area
12 画廊 Art Gallery
13 多功能室 Multi-functional Room
14 洗手间 Toilet

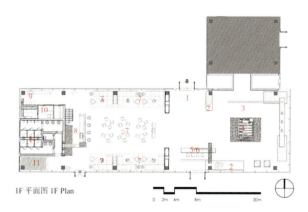

1F 平面图 1F Plan

0 2m 4m   8m   20m

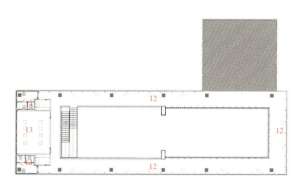

2F 平面图 2F Plan

大厅入口处设有水景，作为室外水景的延伸，模糊内与外的界限。品牌墙直通二楼顶面，纵向延伸了视觉，表面为具有质感的潘多魔涂料，纤细的线条装饰灯则为品牌墙面增添光影的细节。

高达 3 米的接待台背景墙采用团队自主研发的一款天然火山岩"宇宙奇迹"材料，孔洞纹理自然，呈现星光点点的视觉效果，犹如浩瀚宇宙。接待台由石材和金属构成，1 公厘米厚的钢板由台面直插地面，部分台面搭在水波纹大理石上方，呈现出极强的几何构成感。

The entrance to the lobby features a water feature as an extension of the outdoor water feature, blurring the boundary between inside and outside. The branding wall leads to the ceiling of the second floor, extending the view vertically, with a textured Pandemonium paint surface and slim line decorative lights adding light and shadow details to the wall.

The 3m high reception desk backdrop is made from a natural volcanic rock "XY+Z DESIGN", a material developed by the team, with a natural texture of holes, giving a starry visual effect, like the vastness of the universe. The reception desk is made of stone and metal, with a 1cm thick steel plate inserted directly into the floor from the desk top, part of which rests on top of the water ripple marble, giving it a strong sense of geometric composition.

沙盘区采用非常规设计，整个沙盘的底座对角处为石头雕塑，出于安全考虑，计算排布了透明树脂在沙盘底座辅助支撑，沙盘台面为黑色水波纹大理石，让沙盘模型被"水"包围，凝固之水既呼应了当地的运河文化，也暗喻将汴京的辉煌历史凝固在倒影之中。

沙盘区上方整体被"江山"艺术超白玻璃包围，玻璃上的图案为长短不一的线条组成的连贯山形，绵延起伏亦呼应了建筑顶部的曲线结构。图案由设计师精心排布而成，由于玻璃体积较大，安装难度也较大，最大的玻璃单块高度接近8米，在视觉上相当震撼， 同时也成为中央《清明上河图》的最佳背景。两岸重湖叠巘，云树堤沙，中央千骑高牙，吟赏烟霞，两者浑然一体，虚实相交。昔日汴京的盛世繁荣浮现在空中，悠远空灵，将售楼处的主题"上河"生动地呈现在世人眼前。

The sandbox area adopts an unconventional design, with stone sculptures at the diagonal of the base of the sandbox. for safety reasons, transparent resin is calculated and arranged to support the sandbox base, and the table top is made with black water ripple marble, so that the sandbox model is surrounded by "water". the frozen water echoes the local canal culture, and also implies the glorious history of Bianjing.

The sandbox area is surrounded by an artwork made with super white glass, the pattern on which is a continuous mountain shape made up of lines of varying lengths, echoing the curved structure of the top of the building. The pattern is carefully laid out by the designers and is difficult to install due to the size of the glass, with the largest piece of glass being nearly 8 meters high, making it visually stunning and a perfect backdrop for the central painting. The two sides of the river are stacked with clouds, trees and sand, while the central part is filled with a thousand horsemen enjoying the smoke and haze. The prosperity of Bianjing in the past emerges in the air, far away and ethereal, presenting the theme of the sales office "Shanghe" (riverside) vividly in front of the viewers.

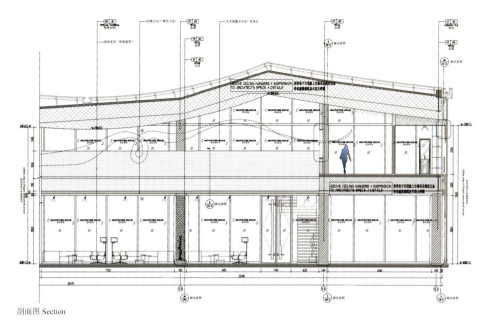

剖面图 Section

房地产示范区复合人文功能空间 COMPOUND HUMANISTIC FUNCTION SPACE OF MARKETING CENTER   067

江艺术馆
ESTUARY ART GALLERY

2021 金堂奖上海赛区年度十佳
2021 入展设计中国全国作品成果展

■

业主单位：广东中天集团 - 中天 · 江湾天地销售中心
景观设计：北京 PRA 建筑设计咨询有限公司上海分公司
建筑设计：上海中森建筑与工程设计顾问有限公司
施工单位：深圳市博业装饰工程有限公司
项目地点：江西南昌
面积：1517 平方米
完工时间：2020

Client: Guangzhou Zhongtian Holding Group/Zhongtian
　　　　 Jiangwantiandi Sales Center
Landscape Design: Beijing PRA Architect (Shanghai Office)
Architectural Design: ZSA Shanghai
Constructor: Boye Decoration Shenzhen
Project Location: Nanchang, Jiangxi Province
Area: 1,517 sqm
Completion Time: 2020

江艺术馆·品鉴中心是我们为南昌中天江湾天地商业中心江艺术馆项目量身打造的集艺术品、展示、会所于一体的高端复合功能型品鉴中心，由主展馆及品鉴中心两栋楼组成，一期为开放艺术品鉴交流中心，二期整体呈现，兼具项目示范区展示功能。设计以滨江天幕为灵感，采取颠覆设计，将艺术搬进生活，让美学质感借由空间体验渗透至生活中的每一个细节。

ESTUARY Art Gallery Discovery Center is a high-end complex center integrating artwork, exhibition and clubhouse, which is designed by the design team for the ESTUARY Art Gallery Discovery Center project in ESTUARY CITY Commercial Center, Nanchang, and consists of two buildings: the main exhibition hall and the Discovery Center. Inspired by the riverside canopy, the design is subversive, bringing art into life and allowing the aesthetic quality to permeate every detail of life through the spatial experience.

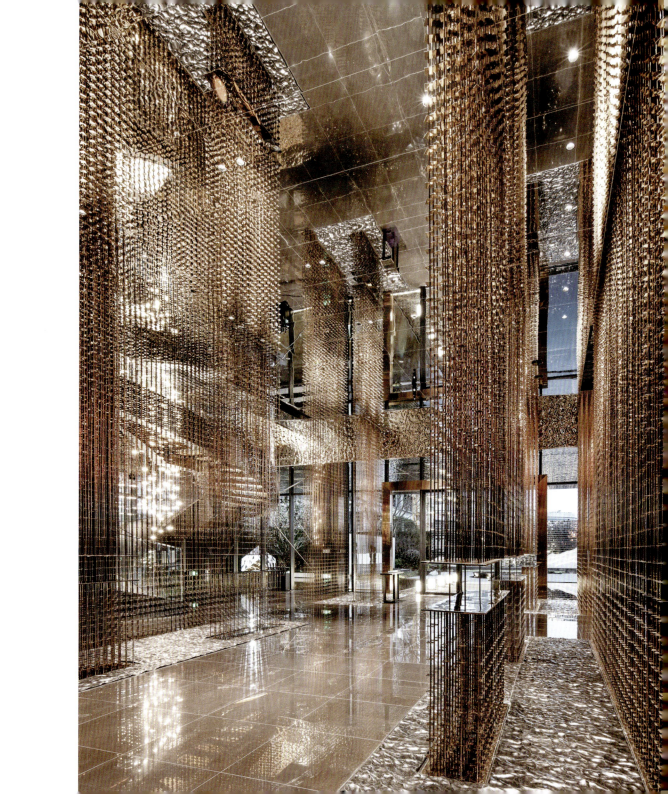

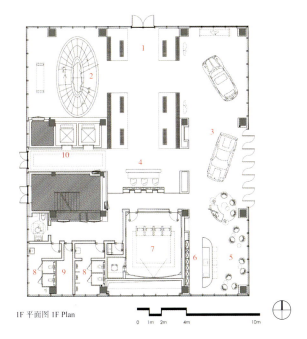

1F 平面图 1F Plan

0 1m 2m 4m   10m

1 售楼部入口 Entrance
2 旋转楼梯 Spiral Stairway
3 展示区 Showcase
4 前台招待 Recetionist
5 休闲区 Leisure Area
6 水吧 Water Bar
7 影音室 Video Room
8 洗手间 Toilet
9 储物室 Storage Room
10 电梯厅 Elevator Hall

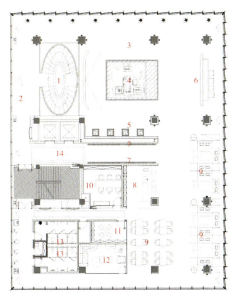

2F 平面图 2F Plan

1 旋转楼梯 Spiral Stairway
2 景观长廊 Landscape Gallery
3 品牌展示区 Brand Display Area
4 大型沙盘 Large Sand Table
5 户型沙盘 Model Sand Table
6 销控台 Sales Control Station
7 金属艺术装置 Metal Art Device
8 水吧 Water Bar
9 洽谈区 Negotiation Area
10 签约室 Signing Room
11 儿童娱乐区 Children's Area
12 贵宾室 VIP Room
13 洗手间 Toilet
14 电梯厅 Elevator Hall

房地产示范区复合人文功能空间 COMPOUND HUMANISTIC FUNCTION SPACE OF MARKETING CENTER 073

空间布局自下而上由公共开放逐渐过渡至高端私密的专属空间：首层艺术展厅，二层楼盘展示销售，三层贵宾接待及体验区。9 米高的"云梯"贯穿三层垂直空间，二层挑高 6 米的沙盘展示区与有如悬浮其上的夹层，丰富了整个建筑纵向空间的互动与交流。滨江天幕式的设计手法，打造串联历史、现在与未来的文化地标，留下让南昌人看得见过去、想得到未来的城市生命脉络。概念上从个性、旅程与记忆三个关键展开并贯穿其中，希望创造一个极具吸引力的空间来满足访者的所有感官，将空间的体验打造成一个层次丰富、无限回味的旅程。

The spatial layout gradually transitions from public openness to a high-end private and exclusive space from the bottom up: an art exhibition hall on the first floor, a property showcase and sales area on the second floor, and a VIP reception and experience area on the third floor. a 9-meter-high "cloud ladder" runs through the vertical space on the three floors, and a 6-meter-high sandbox display area and a suspended mezzanine on the second floor enrich the vertical space of the building. The design is inspired by the riverside canopy, creating a cultural landmark that connects history, present and future, leaving behind a city life cycle that allows Nanchangers to see the past and think about the future. The concept is based on the three keys of personality, journey and memory, and is intended to create an attractive space that satisfies all of the visitor's senses, making the experience of the space a richly layered and endlessly evocative journey.

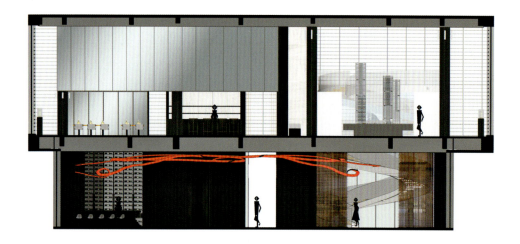

立面图 Elevations

室内景观以金属元素为主，置入由镜面包裹的结构，制造并增强了无限延伸的空间感。地面 3D 水纹雕刻黑金沙大理石营造出水波般的流动感，与精细优雅的电镀不锈钢结构相得益彰，成为窗外滨江景色的延伸，弱化了室内与室外的界限。水波之上，镜面不锈钢艺术装置反射上下倒影串联起关于南昌的一个个记忆碎片，六万片记忆融合成一座空域。在光的作用下渲染出波光粼粼的迷人变幻效果，自然地成为空间的引导者。矩阵式的排列呈现出空间的秩序之美，在若隐若现中将空间划分出相应的功能区。

The predominantly metallic interior landscape incorporates a mirrored structure that creates and enhances a sense of infinite spatial extension. The 3D water-sculpted black sands marble on the floor creates a sense of flowing water waves, complementing the fine and elegant plated stainless steel structure, which becomes an extension of the riverfront view outside the window, weakening the boundary between the interior and the exterior. On top of the water waves, the mirrored stainless steel art installation reflects up and down linking fragments of memory of Nanchang; 60,000 pieces of memory fused into one empty space. The light is a fascinating variation of shimmering waves, which naturally become the guide of the space. The matrix arrangement presents the beauty of order in the space, dividing the space into corresponding functional areas in a hidden way.

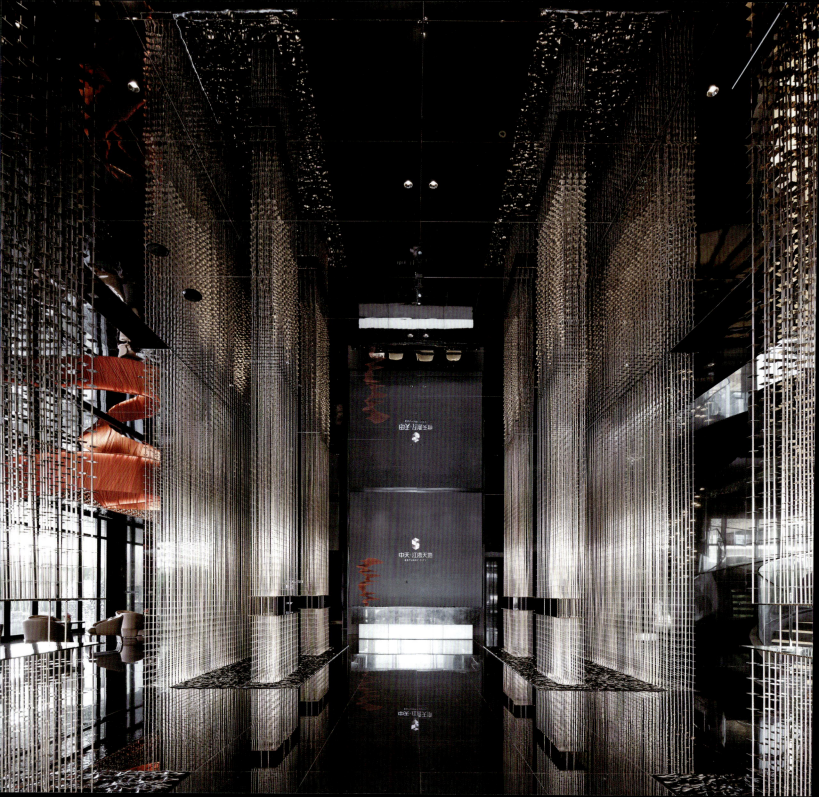

镜面不锈钢顶面与大幅落地玻璃立面令水平及纵向空间同时得到延伸，不仅在视觉上极大提升了空间的尺度感，同时也促成了空间内外、上下、虚实之间的变换，形成独特的感官体验。临江一侧的生活体验区，大画幅玻璃立面将一线江景尽收眼底，上方艺术装置灵感来自当地传统民间技艺糖画与南昌这座城市的红色革命基因。镜面瓷砖与水波纹不锈钢模糊了室内外边界，让空间得到无限延伸。"云梯"串联起整个空间的纵向连接，线条简洁流畅，如行云流水为原本方正硬朗的空间注入恰如其分的柔情。设计将珍贵的赣江历史影像微缩刻入悬垂下来的装饰灯，成为赣江记忆的缩影，与今日窗外的赣江实景同时呈现在一个空间里。

The mirrored stainless steel roof and large floor-to-ceiling glass façade extend both the horizontal and vertical space, not only visually enhancing the sense of scale of the space, but also contributing to the transformation of the space between inside and outside, above and below, and between reality and illusion, creating a unique sensory experience. The living experience area on the riverside features a large glass façade with a panoramic view of the river, while the art installation above is inspired by the traditional local folk art of sugar painting and the red revolutionary DNA of Nanchang. Mirrored tiles and corrugated stainless steel blur the boundary between interior and exterior, allowing for a wireless extension of space. The "ladder" connects the entire space vertically, with simple and smooth lines that flow like clouds and water, injecting the right amount of tenderness into an otherwise square and rigid space. The design incorporates a miniature image of the local history into the hanging decorative lamp, which epitomizes the local memory and is presented in the space alongside the actual scenery outside the window today.

深邃的黑色从地面延续到错落有致的立柱及艺术品展示台，精心设计的排列与沙盘建筑模型共同构成空间秩序的仪式感。热熔玉石玻璃背景墙独特的质感肌理如江水般自然流畅、细腻的色彩、独特的工艺，完美契合品牌的美学内涵，成为品位与尊贵的体现，散发出玉石般温润的光彩，令沙盘区成为整个空间的视觉焦点。同层洽谈区延续了低调而尊贵的黑金色调，柔和的灯光、散落的沙发、进口牛皮搭配黄铜，浓缩了生活美学的内涵与品味，低调简约舒适。

The deep black color continues from the ground to the staggered columns and artwork showcase, and the carefully designed arrangement and the sand building model together form a ritual sense of space order. The unique texture of the hot fused jade glass background wall is as natural and smooth as river water, with delicate colors and unique craftsmanship, which perfectly fits the aesthetic connotation of the brand and becomes the embodiment of taste and dignity, emitting a jade-like warm glow, making the sand table area the visual focus of the whole space. on the same floor, the negotiation area continues the low-key and dignified black and gold tones: soft lighting, scattered sofas, imported cowhide with brass, showcasing low-key simplicity and comfort.

夹层区域的波浪盒子悬浮在二楼上方，茶色玻璃与建筑形式呼应形成记忆点，建立起高低空间之间的互动。低处顶面采用不锈钢镜面，深色主调搭配柔和的灯光，朦胧了虚实交汇的界限，不仅在视觉上提升空间的尺度感，令空间更显大气，同时也丰富了空间的感官体验。

The wave box in the mezzanine area is suspended above the second floor, and the teal glass echoes the architectural form to form a memory point, forming an interaction between high and low spaces. The stainless steel mirror is used on the top surface of the low area, and the dark main color with soft lighting blurs the boundary between the intersection of reality and illusion, which not only visually enhances the sense of scale, but also enriches the sensory experience of the space.

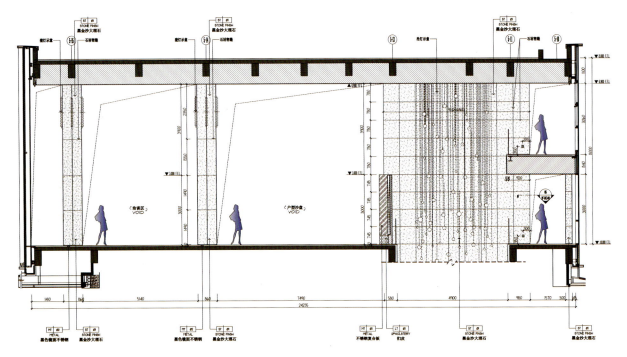

立面图 Elevation

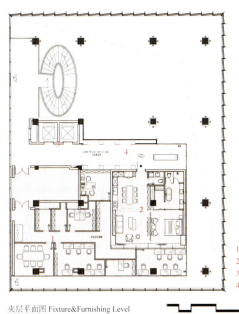

1 办公区 Office Area
2 样板间 Demonstration Room
3 电梯厅 Elevator Hall
4 工艺做法展示区 Craft Techniques Display Area

夹层平面图 Fixture&Furnishing Level

0 1m 2m   4m   10m

3F 平面图 3F Plan

1 多功能室 Multi-functional Room
2 屋顶景观 Roof View
3 样板间 Demonstration Room
4 电梯厅 Elevator Hall

长乐诗词馆

CHANGYUE POETRY MUSEUM

■

业主单位：西安泽登置业有限公司 - 朗诗西安南姜销售中心
景观设计：西安枚犁景观设计咨询有限公司
建筑设计：重庆长厦安基建筑设计有限公司
施工单位：西安大明宫装饰工程有限公司
项目地点：陕西西安
面积：890 平方米
完工时间：2021

Client: Xi'an Zedeng Real Estate Co./Landsea Nan Jiang Sales Center, Xi'An
Landscape Design: Xi'an Jiuli Landscape Design & Consulting Co.
Architectural Design: Chongqing Changsha Anji Architectural Design Co.
Constructor: Xi'an Daminggong Decoration Engineering Co.
Project Location: Xi'an, Shaanxi Province
Area: 890 sqm
Completion Time: 2021

朗诗西安南姜销售中心作为 XY+Z DESIGN 与朗诗集团联手为西安这座历史悠久的城市打造的复合功能型销售中心，集沉浸文化、艺术展示、休闲娱乐、绿色人居为一体。我们旨在传承中华诗词文化的艺术境界、延展西安古都韵味的地域文化，用现代、艺术、创新、独特的设计手法，呈现西安追求品质生活进阶的诉求，打造未来城市居住环境的理想范本。

The Landsea Nanjiang Sales Center is a composite sales center jointly built by XY+Z DESIGN and Landsea group for Xi'an, a city with a long history. it integrates immersive culture, art exhibition, leisure and entertainment, and green residence. The designer aims at inheriting the artistic realm of Chinese poetry culture and extending the local culture of Xi'an as an ancient capital. With modern, artistic, innovative and unique design techniques, the designer presents Xi'an's pursuit of quality life and realizes the ideal model of future urban living environment.

项目坐落于今西安市雁塔区南姜村，曾是羌族一支聚居之地，"羌"字源自甲骨文，意指少数民族，后慢慢演化成汉字"姜"用作姓氏，至唐朝，百姓便唤此地为南姜村，并一直沿用至今。今天，昔日的辽族部落正在成为充满活力的高新开发区，正如这座城市，处处焕发着一种独有的古老却又年轻的生命力。

我们将这种生命力注入空间设计中，运用传统元素与现代空间的结合，将这片土地的记忆镌刻在现代品质生活空间体验中，赋予其新的生命力。

The project is located in Nanjiang Village, Yanta District, Xi'an, which was once home to a group of the Qiang tribe. The word "Qiang" originated from the oracle bones, meaning minority, and later evolved into the Chinese character "Qiang", which was used as a family name. In the Tang Dynasty, the people called this place Nanjiang Village and it has been used ever since. Today, the former Liao tribe is becoming a vibrant high-tech development area, just like the city, which has a unique ancient yet young vitality everywhere.

The design team has injected this vitality into the design of the space, using combination of traditional elements and modern space to engrave the memory of this land into a modern quality living space, giving it a new life.

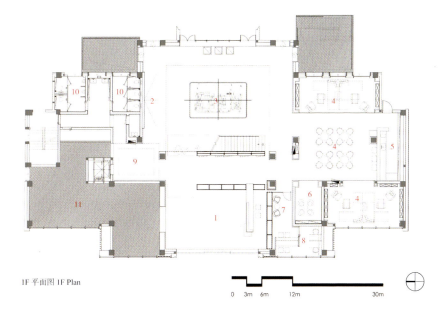

1 大堂 Lobby
2 LED 屏幕 LED Screen
3 大型沙盘 Large Sand Table
4 洽谈区 Negotiation Area
5 水吧 Water Bar
6 儿童娱乐区 Chirdren's Area
7 收银 Cashier
8 财务办公室 Financial Office
9 电梯厅 Elevator Hall
10 洗手间 Toilet
11 品牌馆 Brand Pavilion

1F 平面图 1F Plan

0 3m 6m   12m   30m

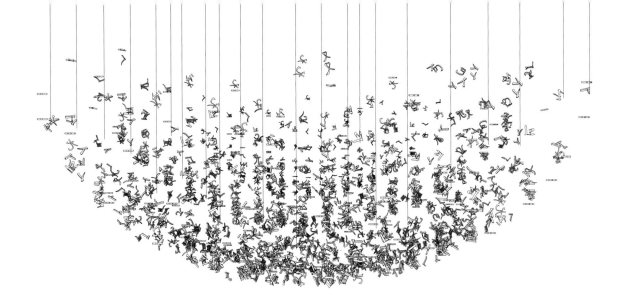

以中华源远流长的诗词文化作为传承西安城市精神的艺术象征。《过华清宫》《客中行》《和贾舍人早朝大明宫之作》作为长安盛世的代表，以此作为空间与城市的融合点，经取形、提色、绘意，从中提取出相应的汉字元素，以砍体、颜体、柳体为形，将汉字、诗词、书法三大中华文明的绚丽瑰宝合为一体，通过现代金属打印技术，实现字、形、意、色合一的艺术设计。并将其运用到空间设计之中，千年的文明因现代设计而以更加艺术的形式在空间中焕发生机。

千"弦"万"语"，三千丝弦串联起上下五千年的中华文明，也串联起整个空间的情感链接，通过现代艺术手法采用令人意想不到的方式使中华民族的灿烂瑰宝再次绽放出举世无双的夺目光彩，为整个空间注入人文情怀的灵魂，成为空间最大的亮点。

千言万语，漫天诗书，诉不尽长安千秋万载，沧海桑田。

The long-standing Chinese poetic culture is used as an artistic symbol to pass on the spirit of the city of Xi'an. The "Passing through the Huaqing Palace", "The Journey of the Guest" and "The Morning Pilgrimage to the Daming Palace with Jia zhi", as the representatives of Chang'an's prosperity, are used as the integration between the space and the city, and the corresponding elements of Chinese characters are extracted from them by taking the shape, color and meaning, and using the chopping style, Yan style and Liu style as the form, the three glorious treasures of Chinese civilization (Chinese characters, poetry and calligraphy)are combined into one. The design is a combination of character, form, meaning and color. The design of this piece of art has been applied to the design of the space, and the thousand-year-old civilization is brought to life in the space in a more artistic form.

The three thousand strings link the Chinese civilization of the last five thousand years and also acts as an emotional link in the whole space, making the splendid treasures of the Chinese nation blossom again in an unexpected way. The installation is also linked to the whole space. It is the most important highlight of the space, injecting the soul of humanism into the whole space.

A thousand words or a thousand poems, cannot tell the story of Chang'an's thousands of years.

室内各功能区依照开放式的建筑语言和行走动线而设，展示、洽谈、生活体验等公共区域实现一体化的流畅衔接，以当代美学格调的构图与铺陈，凸显品质生活的意境之美。

售楼处首层融合展示、洽谈及酒吧于一体。二层则拥有独立书吧、多功能宴会厅等生活体验区域。各区域功能明确，相对独立的同时亦具有丰富的拓展性和可能性。

前厅的几何造型及体块的穿插方式运用了建筑的构筑手法，借鉴中式园林点、障、透的手法，进行淡韵清雅的场景演绎。从玻璃装置与空隙中可眺望大堂中部沙盘区，与一览无余的直白相比，平添几分"犹抱琵琶半遮面"的含蓄与神秘，令人更欲一探究竟，赋予空间以古代园林"借景"的趣味，增强了人与空间、空间与空间之间的互动。

The interior functional areas are designed in accordance with the open architectural language and walking lines, and the public areas such as display, negotiation and living experience are integrated and smoothly connected, with contemporary aesthetic composition and layout, highlighting the beauty of quality life.

The first floor of the sales office is a combination of exhibition, negotiation and bar. On the second floor, there is a separate book bar, a multi-functional banquet hall and other living experience areas. Each area has a clear function and is relatively independent, but at the same time has a wealth of expansion and possibilities.

The geometry of the front hall and the interplay of the blocks use architectural construction techniques. The lightness and elegance of the scene is interpreted by drawing on the techniques of pointing, blocking and penetration in Chinese gardens. From the glass installation and gaps, one can look into the sandbox area in the middle of the lobby, adding a bit of subtlety and mystery compared to the straightforwardness of a glance, making people want to explore more.

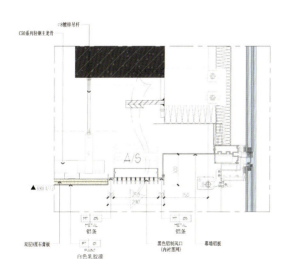

节点大样图 Detailed Drawing

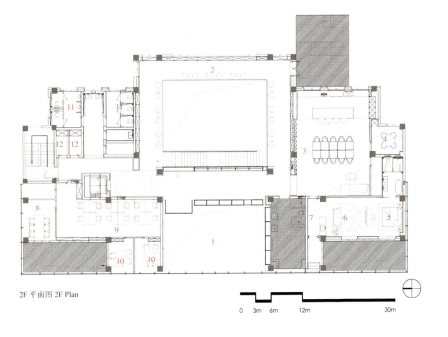

2F 平面图 2F Plan

0 3m 6m   12m   30m

而文字艺术装置"阅读者",其色彩来自古画中的朱砂红,结合现代科技3D 打印技术,与顶面"千弦万语"相互感知、呼应,既是现代的,也是传统的,成为跨越时空的亮丽风景。

西安自古至今都是全国酒文化的核心发展区域。"取酒虾蟆陵下,家家守岁传卮"的酒文化传承至今。故而开启了"长乐诗词馆"的另一重身份——"长乐诗词吧"。古时在长安的饮酒作诗,可在今日的沉浸酒吧中再现。

The installation "The Reader", in the color of cinnabar red from ancient paintings, combined with modern technology (3D printing), and the top surface of a thousand strings, echoing each other, is both modern and traditional, becoming a beautiful landscape across time and space.

Xi'an has been the core development area of wine culture in China since ancient times, and the wine culture of "taking wine from under the shrimp toad's tomb and passing on the goblet to every family" has been passed down to this day. This is why Changyue Poetry Museum is also known as "Changyue Poetry Bar", where the ancient drinking and poetry in Chang'an can be recreated in today's immersive bar.

展示区中，大型艺术装置"文字之源"的背景，我们采用了高达 7 米的发光玉石玻璃，天然玉石温润遒劲，肌理自然隐约可见，铺垫出墨色山水般的空间基调。其特有的纹理、光泽和历史感如同一块巨大的石碑，承载起中华文明之源，正如西安这片土地亦是承载中国民族文明发源的底片。

我们运用立体构成、对比、借景等艺术手法，在明暗、虚实、传统与现代的对立与统一中呈现深邃的东方哲学智慧，通过传统文化元素与现代技术的融合，用艺术空间全新诠释长安这座古都的绝世风采，以史为底、以诗入画、以字入魂，气韵悠长，令人过目难忘。

The display area uses a 7m high luminous jade glass as the natural backdrop for the large art installation "Origin of Words", with the warmth and strength of natural jade and its natural texture visible, laying the tone of the space like an ink landscape. Its unique texture, luster and sense of history are like a giant stone monument, carrying the source of Chinese civilization, just as the land of Xi'an is also the negative film carrying the origin of Chinese civilization.

The design team uses three-dimensional composition, contrast and borrowed scenery to present profound oriental philosophical wisdom in the opposition and unity of light and dark, reality and illusion, tradition and modernity, and through the fusion of traditional cultural elements and modern technology, a new interpretation of the ancient capital of Chang'an with artistic space, using history as the base, poetry as the painting and words as the soul, with a long and unforgettable rhythm.

节点大样图 Detailed Drawing

洽谈区结合酒吧的功能属性，以年轻人群为导向，营造更具活力、舒适、轻松、时尚的体验氛围。设计以酒中漂浮的气沫为灵感，将其不断上升、漂浮的形态，利用艺术化的缩影作为空间的设计语言，融入空间的装饰中，为空间注入源源不断的流动感与趣味化。

The negotiation area combines the functional attributes of a bar with a youth-oriented atmosphere to create a more dynamic, comfortable, relaxed and stylish experience. The design is inspired by the floating air foam in the wine, its rising and floating form, artistically epitomized as the design language of the space and integrated into the decoration of the space, injecting a sense of constant flow and interest into the space.

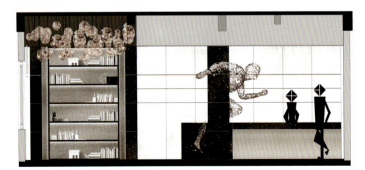

立面图 Elevation

朗诗集团绿色、进取、人本、阳光的核心精神与西安新型人居环境的诉求相结合，我们为洽谈区专属定制的艺术装置"奔跑者"应此而生。正如朗诗所言，美好的生活在于我们不断进取，美好的居住环境更是绿色阳光的，设计更是以人为本的。

本次项目设计中，艺术与设计得以高度融合，为展现最佳效果，我们为项目度身定制了多项专项设计，并取得相关专利。除大型汉字装置"文字之源"、文字雕塑"阅读者"与"奔跑者"外，还包括二层过道的绿色浮岛，大堂中以泡沫切面形状为设计元素的户型台装置——"融合"，以及微景观茶几——"忆唐"。将艺术、绿色、生活进行融合创新的这一系列作品充分体现了 XY+Z DESIGN 一贯追求的设计理念。

The Landsea Group's core spirit of green, progressive, people-oriented and sunshine is combined with the demands of Xi'an's new living environment, and the design team has created the "Runner", an art installation exclusively made for the negotiation area. As Landsea says, a good life lies in our continuous progress, a good living environment is green and sunny, and design is people-oriented.

In this project, art and design are highly integrated, and the design team has tailored a number of special designs for the project, and obtained many patents. In addition to the large Chinese character installation "origin of words" and the sculptures "Readers" and "Runners", the project includes a green floating island in the second floor aisle, a display table in the lobby with shapes of cut-out foams as a design element, and a coffee table as a micro landscape. This series of works, which integrates art, green and life, fully embodies the design philosophy that XY+Z DESIGN has always pursued.

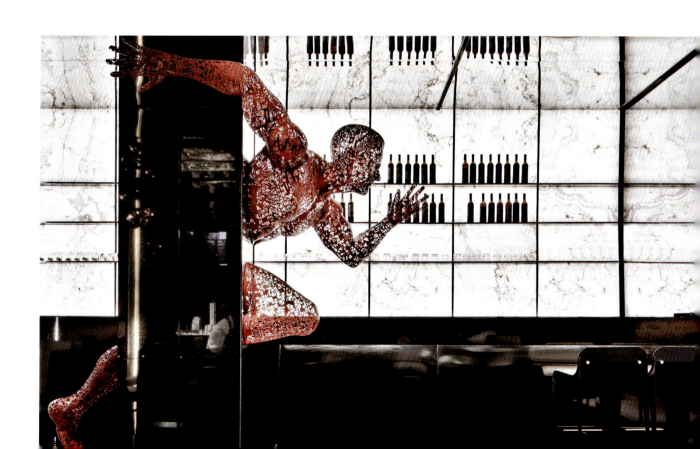

云端
THE CLOUD

黑森林智能实验室
DARK FOREST INTELLIGENCE LAB

源数据粒子空间站
SOURCE DATA PARTICLE SPACE STATION

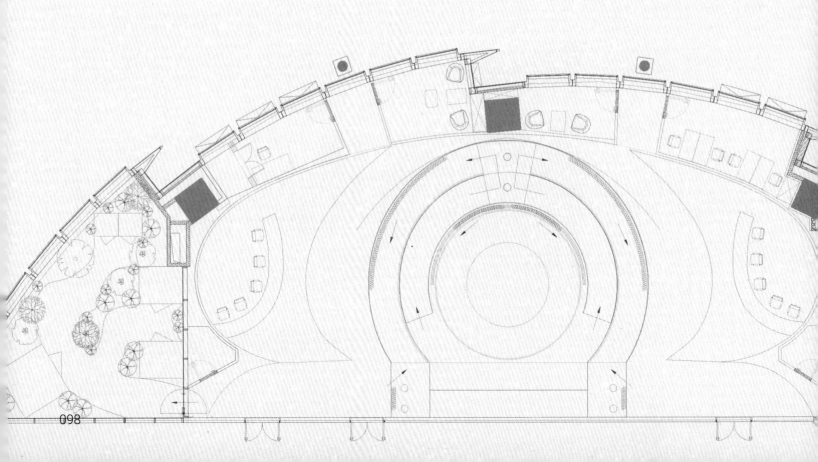

展示与智能空间的融合

INTEGRATION OF DISPLAY AND INTELLIGENT SPACE

这里要感谢未来研究院院长对我的认可及支持。多年来,一直负责未来研究院的诸多智能实验室设计。从第一个中国微电影与云创意实验室"云端"开始,仿佛让我打开了与以往设计项目完全不同的未来之门。原来,设计不止是设计,不仅要结合地域文化及项目定位,还需要对非常多的内容进行熟知与理解。比如"科学""科技""智能"。从"云端"到"黑森林智能实验室",再到"苏州实验室",在院长与业主方的支持下,与诸多跨领域的科学家、艺术家、技术人员合作,这些过程都成为我设计生涯的宝贵经历及财富。它们不仅在功能上实现了复合化,更将技术、艺术、科技进行了很好的融合。

I would like to express my gratitude to the Director of the Institute for Future Studies for recognizing and supporting me. Over the years, I have been responsible for designing intelligence laboratories for the Institute. Starting with the first Chinese micro-film and cloud creativity laboratory, "Cloud", it feels as though we have opened a door that leads to the future, distinct from previous design projects. We have come to understand that it is more than just design; it involves a thorough knowledge and understanding of many different aspects, such as "science", "technology", and "intelligence". From "Cloud" to the "Black Forest Lab" and to the "Suzhou Lab", with the support of the Director and the clients, we have collaborated with numerous cross-disciplinary scientists, artists, and technicians, and have incorporated their input into our designs. These experiences have become invaluable treasures in our design career, as they not only achieved functional compounding but also harmoniously united technology, art, and science.

云端
THE CLOUD

2013 "永隆·星空间杯"
　　　江苏省室内设计、陈设设计大奖赛公共空间工程类银奖
2013 中国室内设计大奖赛入选奖
2013 金堂奖公共空间设计年度十佳
2019A'Design Award 设计大奖铜奖

■

业主单位: 南京大学金陵学院 - 微电影与媒体创意实验室
项目地点: 江苏南京
面积: 700 平方米
完工时间: 2012
设计材料: GRG . 钢化玻璃、玻璃钢

Client: Jinling College, Nanjing University/Microfilm and Media
　　　Creativity Lab
Project Location: Nanjing, Jiangsu Province
Area: 700 sqm
Completion Time: 2012
Materials: GRG, tempered glass, FRP

"云端"实验室在悬念和争议声中悄然呈现在蓝天白云之间,落成于金陵学院传媒学院顶层的寂寥天台之上。她不仅让一座扇形不规则天台不再落寞,更让一众天之骄子的无限创意从此自云端涌现。

在云端,实验室不再是枯燥死板的电脑鼠标,它们灵动闪耀,成就一片智慧场域;师生不再是教条般的对立,他们相伴同习,成长与共。

光、水、雾、你、我,在这里融合;技术、艺术、智慧、情感,在这里相遇;过去、现在和未来,在云端永恒。

The "Cloud" laboratory was quietly presented in the blue sky and the white clouds amidst suspense and controversy. it was completed on the lonely rooftop of the School of Communication, Jinling College. It is not only a fan-shaped irregular rooftop that is no longer lonely, but also a place where a lot of creative ideas are coming out of the clouds.

In the clouds, the labs are no longer boring about rigid computers and mouses, but a field of wisdom, where teachers and students are no longer in dogmatic confrontation; they are learning and growing together.

Light, water, fog, you and me, all merge here; technology, art, wisdom and emotion meet here; the past, present and future are eternal in the cloud.

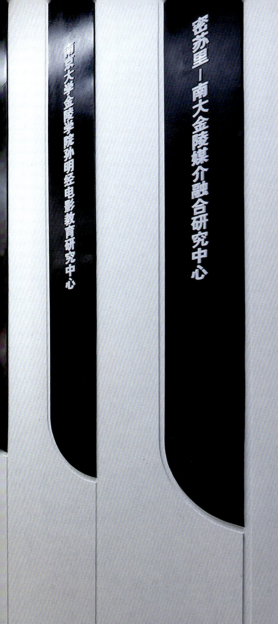

密苏里—南大金陵媒介融合研究中心

南京大学金陵学院孙明经电影教育研究中心

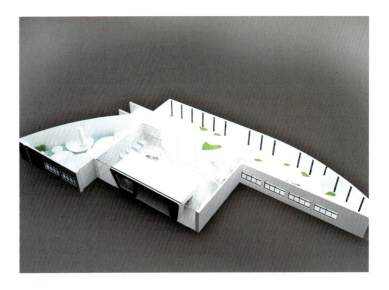

她是自由的：创意区、彩排区、表演区与裸眼 3D 实验室、微博实验室、行为观察室、浮岛演播区之间相互通透、自由流动；同时满足各功能区独立、属性特征明显。两大区域间相互呼应、相互融合。

她是经典的：这里承载着微电影奠基人、民国时期一代电影大师孙明经的胶片梦想，通过调光影像大屏得以一一展现。

She is free: the creative area, the rehearsal area, the performance area and the 3D lab, the microblog lab, the behavioral observation room and the floating is-land performance area are permeable to each other and flow freely; at the same time the functional areas are independent with their distinctive attributes. The two main areas echo and integrate with each other.

She is classic: here are the dreams of Mingjing Sun, founder of micro-film and one of the film masters in the Republic of China. Here, the dimming image screen shows them all.

1 媒体实验室 Media Lab
2 展台（立体投射）Booth (3D Projection)
3 创意实验室 Creativity Lab
4 微博实验室 Microblog Lab
5 传媒情报研究分析中心 Center for Research and Analysis of Media Intelligence
6 室外演播厅 Outdoor Studio
7 集水池 Pool
8 行为观察 Behavioral Observation
9 休闲区 Relaxation Area
10 台阶观演台 Terraced Viewing Platform
11 Logo 墙 Logo Wall

平面图 Plan

0 2m 4m   8m   20m

她是生态的：生态绿化、自然采光、高效通风、资源再生节能（自循环雨水收集净化系统与空气负离子再造系统）……浸润着设计师对生命的敬重，对自然的尊重。

她是科技的：全景电脑灯、智能科技控制、全息投影、智能化搜索、数据挖掘与分析……时时处处挑战着人类对技术进步的认知，彰显着年轻一代对前沿科技的梦想和持续探寻。

She is ecological: ecological greening, natural lighting, efficient ventilation, energy saving by resource regeneration (self-circulating rainwater collection and purification system and negative air ion re-creation system)... it shows the designer's respect for life and for nature.

She is technological: panoramic computer lights, intelligent control, holographic projection, intelligent search, data mining and analysis... it challenges mankind's knowledge, searching for technological progress and dreams of cutting-edge technology at all times.

黑森林智能实验室
DARK FOREST INTELLIGENCE LAB

2019 A'Design Award 设计大奖银奖
2020 金堂奖展示空间设计年度最佳
2019 中国室内设计大奖赛入选奖
2020 入选中国室内设计年鉴

■

业主单位：新华网融媒体未来研究院、如皋市经济技术开发区 - 长
三角车媒体传感智能实验室
项目地点：江苏南通如皋
面积：600 平方米
完工时间：2018
设计材料：布帘、吸音板、PVC 地垫

Client: Future Media Convergence Institute (FMCI), government
of Rugao Economic and Technological Development Zone/
Yangtze River Delta Vehicle Media Sensing Intelligence Lab
Project Location: Rugao, Nantong, Jiangsu Province
Area: 600 sqm
Completion Time: 2018
Materials: Fabric curtains, sound absorbing panels, PVC floor mats

人类文明在未知、感知、认知的历程中逐步成长。

黑森林生物车媒体传感实验室是中国首个基于生物智能前沿技术开发的人机交互与情感计算的重点实验室，旨在通过实验室交互游戏了解人的情绪波动与情感世界，进而对人工智能领域中的生物智能传感研究开启一扇洞察之门。

Human civilization has grown gradually through the journey of the unknown, perception and cognition.

The dark Forest intelligence lab is China's first key lab for human-computer interaction and emotional computing based on cutting-edge bio-intelligence technology. It aims to understand the world of human mood and emotions through interactive games in the lab, thus opening a door of insight into bio-intelligence sensing research in the field of artificial intelligence.

在脑科学与神经科学领域，人类未知的神秘世界被喻为 "黑森林"。我们巧妙借用此概念，用纯黑色营造环境，"蓝星球"的穹顶将宇宙广袤无际置于森林中央，使空间无限次被放大。时空的变化将宇宙与人类命运相连接，顿生壮阔之感。

我们借助空间传递对自然的保护，以"自然生长"为理念，空间注重复合功能，实现 70% 以上的装配化率，80% 以上的可回收利用材料占比，最大化利用空间，满足不同场景变幻需求。

In the field of brain science and neuroscience, the unknown and mysterious world of human beings is referred to as the "dark forest". The designer has cleverly borrowed this concept by creating an environment in pure black, with the "Blue Planet" dome placing the vastness of the universe in the middle of the forest, making the space infinitely magnified. The change in time and space connects the universe with the destiny of mankind, creating a sense of grandeur.

The design uses the space to convey the protection of nature, with the concept of "natural growth". The space focuses on composite functions, achieving an assembly rate of over 70% and a recyclable material ratio of over 80%, maximizing the use of space and meeting the changing needs of different scenes.

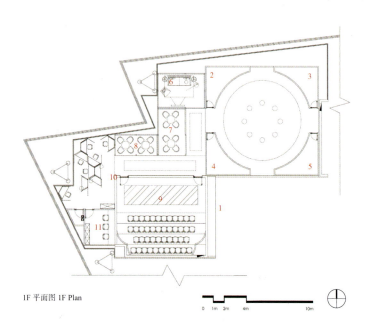

1F 平面图 1F Plan

0 1m 2m 4m 10m

1 序厅 Lobby
2 备用间 1 Spare Room 1
3 备用间 2 Spare Room 2
4 备用间 3 Spare Room 3
5 备用间 4 Spare Room 4
6 访谈室 Interview Room
7 交流区 1 Communication Area 1
8 交流区 2 Communication Area 2
9 影音测评实验室 Audio-Visual Evaluation Laboratory
10 数据处理与分析工作室 Data Processing and Analysis Studio
11 控制室 Monitoring Room

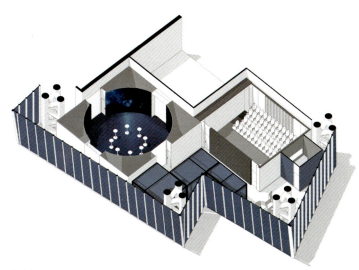

轴测图 Axonometric Drawing

用户在实验室内可随意选择交互方式进行视听体验，包括戒指、手环等穿戴式设备，或者不同材质的、具备生理感应功能的座椅靠垫等。最奇妙的是用户最真实的体验感受被精准记录下来，有效帮助编剧和导演改进作品，也帮助参与实验的用户更深入地了解自己；对于潜在的自闭倾向儿童或失去语言表达能力的痴障老人，也可以早期发现或预防、预警。

从人类的起源、文明的产生到科技的发展，再到爱……"黑森林"可以演绎出全球的文明发展史，以此来探讨人、自然与科技的关系。

Users can freely choose their way of interaction for audio-visual experience in the laboratory, including wearable devices such as rings and bracelets, or seat cushions in different materials with physiological sensing functions. The most wonderful thing is that the user's authentic experience is accurately recorded, effectively helping writers and directors improve their works, and also helping users participating in the experiment to have a deeper understanding of themselves; early detection, prevention, and warning can also be provided for children with potential autism tendencies or elderly individuals with language impairments.

From the birth of humanity, the emergence of civilization, to the development of technology, and then to love... The Dark Forest is to interpret the global history of civilization, in order to explore the relationship between humans, nature, and technology.

立面图 Elevation

借鉴森林场景营造环保低碳与全景声静场环境，为生物传感实验塑造出独特的夜景氛围，静逸时一根针落地的声音和方位都清晰可辨，使体验者专注投入场景剧情中，科学家可以准确捕捉到被试者生理指标数据与情绪值的变化。

未来的设计必将面临跨学科、全媒体、无时差的融合挑战。黑森林实验室建设难度大、设计要求高，涉及脑科学、神经科学、心理学、医学、计算机学、无线通信、影视传播等多个学科。我们经过 27 天超极限的调研、沟通、设计与施工奋战，终于在由全球 20 多个国家 1600 余名代表参加的第三届国际氢能与燃料电池汽车大会召开时圆满落地。

Drawing inspiration from forest scenes, the designers created an environmentally friendly, low-carbon, and panoramic soundscape environment, providing a unique night atmosphere for biosensing experiments. Even the sound of a needle landing on the floor is discernible, so users can enjoy the experiment in the immersive scene, and scientists can accurately capture the changes in their physiological and emotional states.

Future design will inevitably face the challenge of interdisciplinary, all-media and instant integration. The Dark Forest lab is a demanding project that requires high-quality design, involving multiple disciplines such as neuroscience, psychology, medicine, computer science, wireless communication, and film and television communication. After 27 days of extreme efforts in research, communication, design and construction, the design team finally met the deadline and the lab was open before the 3rd International Hydrogen and Fuel Cell Vehicle Congress & Exhibition (FCVC 2018), which received more than 1,600 representatives from more than 20 countries around the world.

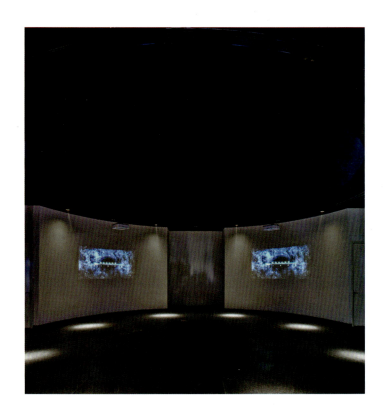

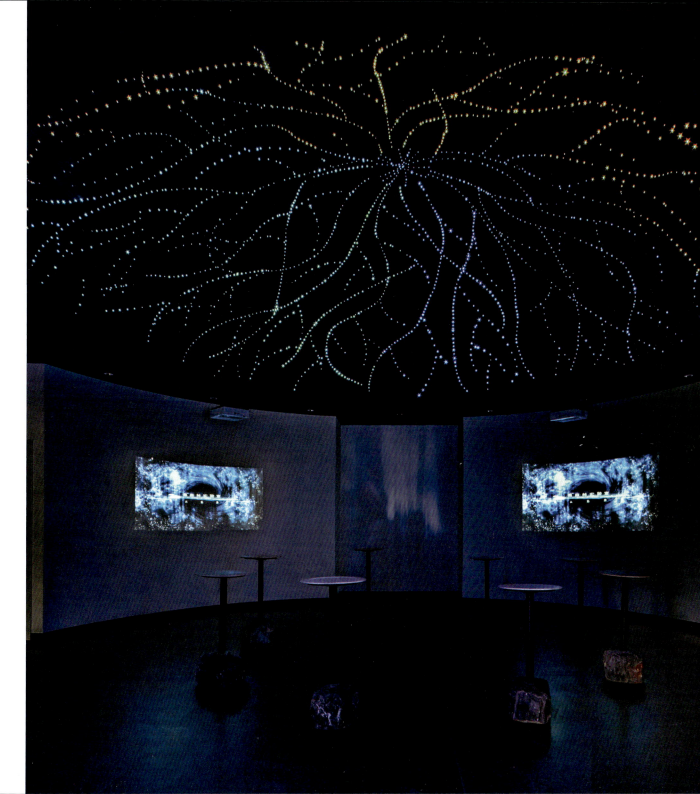

源数据粒子空间站
SOURCE DATA PARTICLE SPACE STATION

2022 美国 IDA 设计大奖 银奖
2022 法国巴黎 DNA 设计大奖 荣誉奖

■

业主单位：苏州日报社 - 苏州日报人机融合智能研发中心
项目规划单位：新华网融媒体未来研究院
项目地点：江苏苏州新闻大厦 25 层
面积：728 平方米
完工时间：2021

Client: Suzhou Daily/Suzhou Daily "Fusion of Human and Machine"
        Intelligent R&D Center
Project Planning Company: Xinhua Net Future Convergence
                          Media Research Institute
Project Location: 25F, News Building, Suzhou, Jiangsu Province
Area: 728 sqm
Completion Time: 2021

源数据是"数据之源"，一切真相的初始与本核。物联网赋予它多类型、异属性、跨领域的复杂性，却又创造人们假以技术、工具、方法、智慧进行识别的能力。

源数据粒子空间站是我们联合新华网、苏州日报打造的融媒体实验探索与创新平台，集遥感、体验、直播、孵化、交互、展示为一体，是一个兼具情感、灵魂与未来的创意空间。

Source data is the "source of data", the beginning and core of all truth. The Internet of Things gives it the complexity of multiple types, heterogeneous attributes and cross-domains, yet creates the ability for people to identify it with technology, tools, methods and wisdom.

The Source Data Particle Space Station is an experimental exploration and innovation platform for integrated media created by the design team in collaboration with Xinhua net and Suzhou Daily, integrating remote sensing, experience, live streaming, incubation, interaction and display; a creative space with emotion, soul and future.

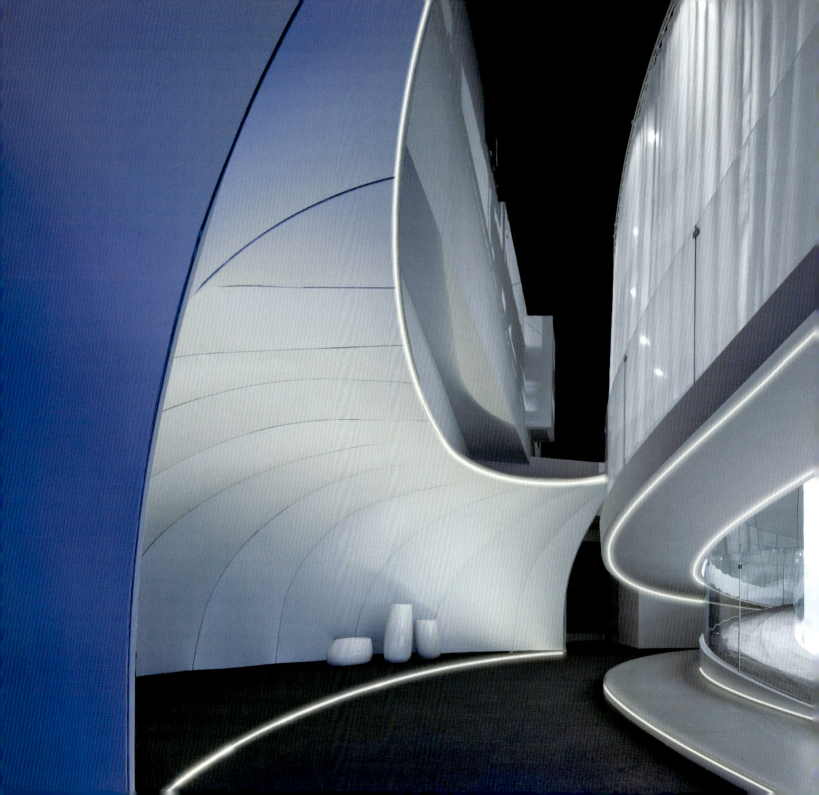

基于对世界多元化和生物多样化的好奇，积淀对宇宙的幻想和未知的敬畏，设计以太阳系行星运行轨迹作为核心概念，以时间与空间的因缘关系，在有限中生无限，融合、流转、变换，循环而永生。

Based on curiosity about the diversity of the world and organisms, and on fantasy and awe towards the universe, the design takes the trajectory of the planets in the solar system as the core concept, trying to interpret the causal relationship between time and space, which involves fusion, transformation, circulation, and immortality.

空间站的未来是科技智能化的，也是绿色可持续的。所展示的内容应是动态的，不断注入新的创意，每次的展览就是一场精心策划的艺术。新颖的展示内容可以引导观众在空间驻留，不同的展示内容，读懂不一样的空间故事，这里是与时空对话的媒介。

空间自有生命，设计赋予灵魂。空间站的灵魂在于永恒流动的曲线，不仅是行星运行轨迹的具体表现，也是设计师赋予空间独一无二的基因。曲线是从自然之美中获取灵感，是一种微妙的引领，将人的视线吸引到不同的时空。行云流水般的曲线向四周延伸、扩展、闭合，周而复始，以影音展现不同的生命感受，沉浸未来的无限畅想。

The future of the space station is green, sustainable and technologically intelligent. The content presented should be dynamic, constantly infused with new ideas, and each exhibition is a carefully curated piece of art. Novel displays can guide the viewer to stay in the space, and this is a medium for dialogue with time and space.

Space has a life of its own and the design gives it a soul. The soul of the space station lies in the eternal flow of the curves, which are not only the formal expression of the planetary orbits, but also the unique DNA that the designers have given to the space. The curves take their inspiration from the beauty of nature and are a subtle guide that draws the eye to a different time and space. The flowing curves extend, expand and close in all directions, over and over again, revealing different sensations of life in audio-visuals and immersing oneself in the infinite imagination of the future.

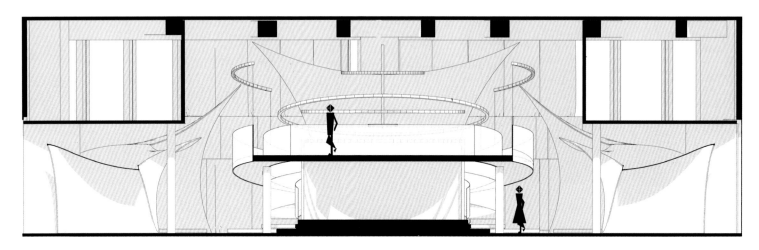

立面图 Elevation

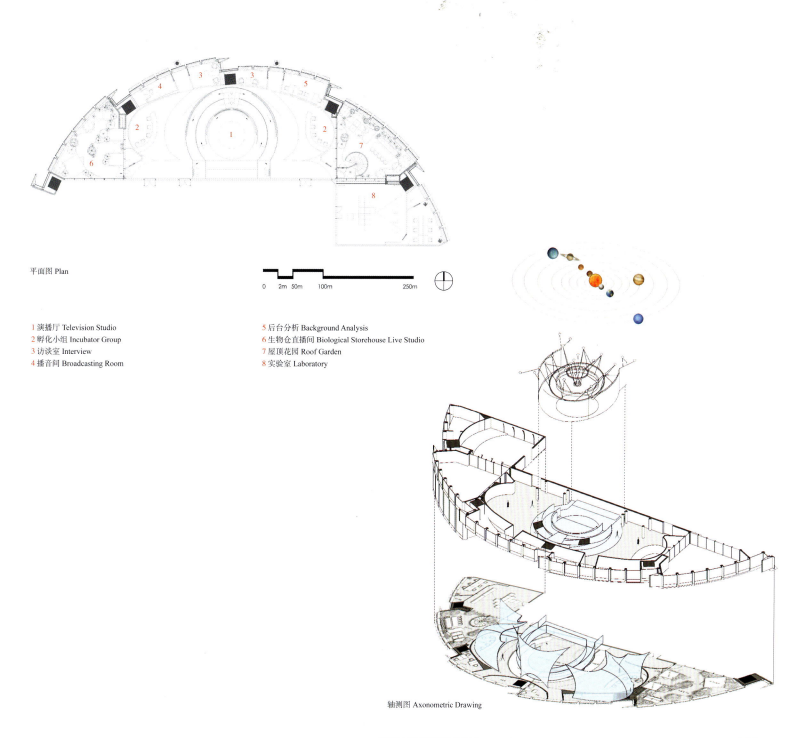

平面图 Plan

1 演播厅 Television Studio
2 孵化小组 Incubator Group
3 访谈室 Interview
4 播音间 Broadcasting Room

5 后台分析 Background Analysis
6 生物仓直播间 Biological Storehouse Live Studio
7 屋顶花园 Roof Garden
8 实验室 Laboratory

轴测图 Axonometric Drawing

展示与智能空间的融合 INTEGRATION OF DISPLAY AND INTELLIGENT SPACE　123

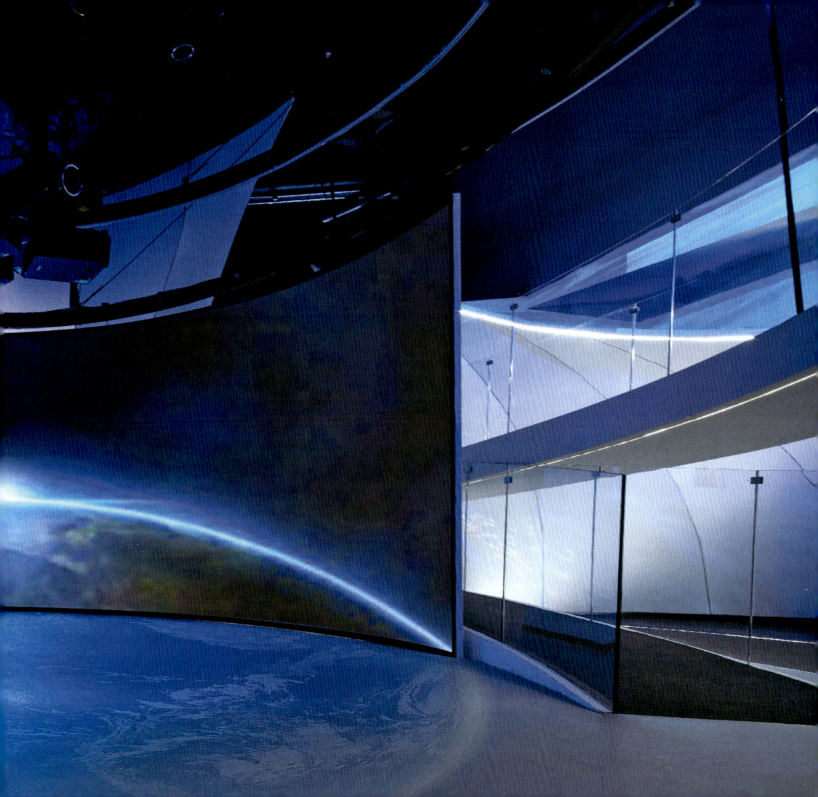

空间站的中心是一个极具设计感与视觉张力的圆形区域，近 8 米高的挑空，是整个空间站的灵魂。区域内设置了一个近半圆的巨型投影屏，外围是电动环形幕布，内外屏幕联动与地面影像构成沉浸式体验；作为整个空间最吸引人的装置，内、外层同步展示，不同的展示内容会呈现迥然不同的空间功能，如"场景还原""电影评测""新闻发布"等，未来在这里可以接纳各种展览、路演、发布会等活动。

The center of the space station is a circular area with visual tension, with a nearly 8m high ceiling, which is the soul of the entire space station. The area is equipped with a giant projection screen in a nearly half circle, with a motorized ring curtain on the periphery, linking the inner and outer screens with the ground images to form an immersive experience; as the most attractive installation in the whole space, the inner and outer layers are displayed simultaneously, with different displays presenting very different spatial functions, such as "scene restoration", "film review" and "press release". In the future, various exhibitions, road shows, press conferences and other activities can be accomodated here.

中心区域上方是模拟行星运行轨迹的圆环装置，两侧分别架起了弧形的浮桥。浮桥是一个弧形双向的通道，由一层旋转向上，外侧被环形幕布包围，立于高处仿佛悬浮于太空之中。浮桥的两侧为融媒体孵化小组，再往外侧是作为办公、录播间、电影播放厅、分析室、设备间等功能区域。浮桥通向的二层仅有一个基于原始结构保留的观察平台，外侧为张拉膜形成的异形半围合空间，将孵化小组包含在内。多重的功能形态为这个空间站注入了灵魂，成为像人一样可以同时承载多重身份的空间，而不是只有一种纯粹、单一的特定功能，它是一个复合化的未来空间。新生与永恒是空间站的未来发展方向，中心区域就像一个破茧新生的蝴蝶：以自由、无限的线条组合形式展现新生，以空间的张力呼应着永恒。

Above the central area is a circular device that simulates planet orbits, with curved floating bridges built on both sides. The bridges are curved two-way passages that rotate upwards, surrounded by a circular curtain, standing high as if suspended in the outer space. On both sides of the floating bridges are places for the integrated media incubation groups, and further to the outside are functional areas such as offices, recording rooms, movie studios, analysis rooms, and equipment rooms. The floating bridges lead to the second floor, where there is only an observation platform, which is based on the existing structure of the building. The outer side is a heteromorphic semi-enclosed space formed by using stretched membrane. Various forms and functions have injected soul into the space station, making it a space that can simultaneously have multiple identities, like a man, rather than just having a single specific function. It is a composite future space. Rebirth and eternity are the future development directions of the space station, and the central area is like a butterfly breaking out of the cocoon: presenting rebirth in a free and infinite combination of lines, echoing eternity with the tension of space.

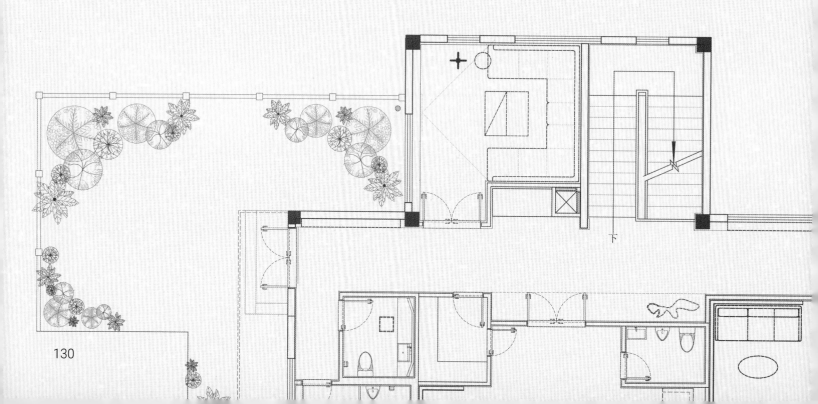

复合性商业空间思考

THOUGHTS OF COMPOUND COMMERCIAL SPACE

复合性空间，我认为更好地表达了我的设计理念。不止设计，不仅是我做多领域的设计，更是让我设计的空间不止是某一种单一功能的空间。这里要再次强调空间的复合生长性。

设计圈经常看到诸如"加班""改方案""苦不堪言"……这样的词汇。设计对有些人而言，就是一份赚钱的工具，也有人觉得这只是家人的期盼。我认为，任何事，都需要建立在"热爱"的基础上，恋人是，工作是，设计更是。如果不爱，自然就苦闷于各种无奈，产生诸多负面情绪。设计对我而言，已经不是单纯的热爱了。设计，是我的氧气，没有它则无法呼吸。如果有一天，我不做设计了，我想，我将活不下去。当下的设计，被切割得太细：住宅空间设计、酒店空间设计、零售空间设计、餐饮空间设计、办公空间设计、茶空间设计、酒空间设计、咖啡空间设计……写到这里，我止不住地皱眉。不可否认，专注于某一类空间设计，一定不太容易失败，因为有太多同类的经验，但随之而来的就是更多重复性的出现。

多年来，常有人问我为什么不专注于某一类空间。我都不敢想象那一天的到来。联想一下，每天只吃同一道菜……生活是丰富多彩的，设计其实也一样。虽说多领域设计相较于专属领域设计而言缺少了相对的经验，但相对的它也会带来跨领域的不同理解和创意。我认为，跨领域设计会带来更意想不到的精彩。

I think the concept of compound space could effectively explain my design philosophy. When I say "beyond design", what I mean is not only I design in different fields, but I design spaces with multiple functions. I'd like to re-emphasize the growth of compound space.

Designers often see words like "overtime", "change plans" and "unbearable suffering"... in their WeChat moments. For some people, design is simply a tool to earn money, while others feel it is fulfilling family expectations. I believe that anything must be built on a "passion", whether it be for a lover, a job or design. Without that love, one will naturally feel burdened and face negative emotions. Design means more to me than mere passion; it is my oxygen, without which I would be unable to breathe. If I ever stopped design work, I think I would not be able to survive. Design work today is divided into too many areas: residential design, hotel design, retail space design, restaurant design, office space design, tea room design, bar design, coffee space design... and I cannot help but frown. While it is undeniable that focusing on a specific area of design makes it less likely for you to fail because there is so much experience available, it also leads to increased repetition.

Over the years, many people have asked me why I do not focus on a specific field of design, but I cannot imagine that day ever coming. It's like eating the same food every day... Life is colorful, and so is design. While it is true that cross-disciplinary design may lack a certain level of expertise compared to focusing only on one field of design, it can bring various new creative perspectives and ideas. I believe that cross-disciplinary design will always bring more surprises than we expect.

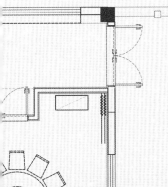

苏州灵美东方医院
SUZHOU LINGMEI ORIENTAL HOSPITAL

运河之滨，太湖之畔，苏州灵美东方医院坐落于此。以水为底，以心为灵，打造优雅色调与舒适空间，让美丽和而不同，赋你万里挑一；以美学点睛灵韵，以医学至臻细节。XY+Z DESIGN 晰纹设计，以灵动融合之势，创造品质现代之美。

苏州灵美东方医院坐落于吴江区的太湖东侧。整体设计是在我们设计的南京灵美东方医院的基础上延续、更新而成的。灵美东方医院是主张东方女性独立自我与和谐之美的医美品牌，始终致力于打造千人千面的国韵之美，坚持从心出发、以心正美的方案准则，希望帮助女性养成真我、自我、本我的独特气质。

■

业主单位：江苏灵美东方医院管理有限公司
项目地点：中国苏州
面积：2400 平方米
完工时间：2022

Client: Jiangsu Lingmei Oriental Hospital Management Co., Ltd.
Project Location: Suzhou, Jiangsu Province
Area: 2,400 sqm
Completion Time: 2022

On the bank of the Canal and the Taihu Lake, Suzhou Lingmei Oriental Hospital is located. The designers at XY+Z DESIGN tried to create elegant tones and comfortable spaces, providing a high-quality place for a modern hospital.

Suzhou Lingmei Oriental Hospital is located to the east of the Taihu Lake in Wujiang District. The design is kind of continuation and update of Nanjing Lingmei Oriental Hospital we designed. Lingmei Oriental Hospital is a medical cosmetology brand that advocates for the beauty of Eastern women. It is always committed to creating the national charm of Chinese women and helping them develop the unique temperament of the oriental beauty.

迈入灵美东方医院的前厅，映入眼帘的是大片的柔和的白色线条细节，以沉稳和明亮之色展现东方女性之美。前厅的曲线融合了水的线条和"凤舞"的主题概念，飞扬的天花板如同浪花，又似凤凰翅膀由两侧向中间回笼，轻柔而包容地笼罩在每个人身上。

Entering the lobby, what catches your eyes are some soft white lines on the ceiling, showcasing the beauty of Eastern women in a calm and bright color. These curved details are inspired by water and combined with the design concept of "phoenix dance". The soaring ceiling looks like waves or phoenix wings.

1F 平面图 | 1F Plan

1 护士站 Nursing Station
2 精神空间 Spiritual Space
3 医生值班室 Doctor's Duty Room
4 观察室 Observation Room
5 咨询室 Consultation Room
6 经理办公室 Manager's Office
7 治疗室 Treatment Room
8 景观庭院 Landscaped Courtyard
9 苏醒室 Awakening Room
10 手术室 Operating Room
11 综合办公 Comprehensive Office
12 牙科诊疗室 Dental Clinic
13 医生办公室 Doctor's Office

0 2m 4m   8m   20m

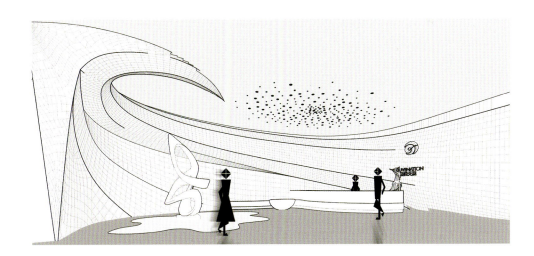

1 吾道吾茶 Tea Room
2 景观露台 Lanscaped Terrace
3 咨询室 Consultation Room
4 VIP 室 VIP Room
5 盖娅展区 Gaia Exhibition Area
6 收银区 Cashier Area
7 吧台 Bar

2F 平面图 2F Plan

0   2m 4m   8m   20m

盖娅传说的服装展厅，它位于苏州分院的二楼中庭，使用大片镜面增强水的反射感并对空间进行拓展，营造优雅而通透的空间，将自然之道融入道法自然的哲学慧思，将生命之美与灵性智慧融于现代设计天人合一的精妙构思。

Gaia, an exhibition area, is located in the atrium on the second floor. Here, large mirrors are used to enhance the reflection of water and expand the room, creating an elegant and transparent space. The design idea is to integrate the way of nature into the philosophical wisdom of Taoism, to communicate the idea of "unity of nature and man".

古之茶经，常与禅意相通。意境之味，即与"吾道吾茶"茶馆相连。灵美东方医院会在每一家医院里都融入茶文化，以强调东方之韵。设计中选择了更加深沉浓郁的棕色及深灰色配色，用以表达禅意东方的经典概念。

The ancient tea culture is often associated with Zen. Lingmei Oriental Hospital tries to integrate tea culture into each of their branches to emphasize the oriental charm. In the design, a color scheme of rich brown and dark gray was chosen to communicate the classic concept of Zen in the East.

3F 平面图 3F Plan

0 1m 2m 4m   10m

1 总经理办公室 General Manager's Room
2 电梯厅 Elevator Hall
3 疏散楼梯 Escape Stairs
4 药房 Pharmacy
5 皮肤治疗室 Skin Treatment
6 注射室 Injection Room
7 输液室 Transfusion Room
8 摄影间 Photography
9 卸妆间 Makeup Remover
10 储藏间 Store Room
11 皮肤科 Dermatology
12 护士站 Nursing Station
13 等待区 Waiting Area
14 财务室 Accounting Office
15 皮肤检测室 Skin Test Room
16 麻醉科 Anesthesiology
17 美容外科 Aesthetic Surgery
18 检验室 Checkout Room

向上走去，便是走向的三楼的楼梯。楼梯间上摆放着巨大的木艺装饰物，以绿色植物点缀旋转蜿蜒的原木色装置，形状犹如凤凰起飞，寓意了循环和再生。

Keep going up, and you will find the stairs on the third floor. There are huge wooden decorations on the stairwell, and the rotating and winding log colored device is decorated with green plants. The shape is like a phoenix taking off, implying circulation and regeneration.

走上三楼就到了诊疗区的护士站，这里采用了直线与弧线搭配的台面及吊顶，这些精妙设计让等待区和护士站不再单调，而成为一个可以休息、打卡的自由活动区域。

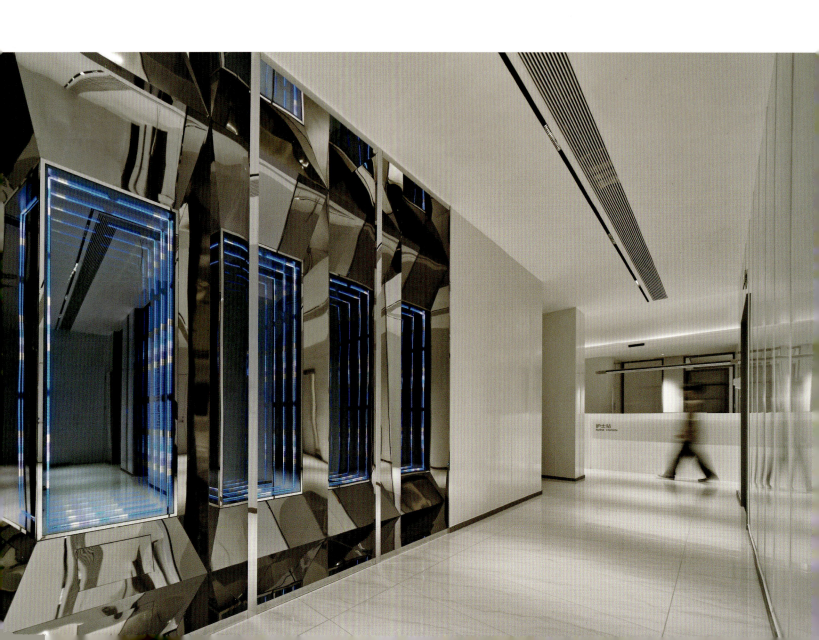

Walk up to the third floor and you will reach the nurse station in the diagnosis and treatment area. The counter top and ceiling with straight lines and arcs are used here, making the waiting area and nurse station no longer monotonous, but a free activity area where you can rest and selfie.

五行学说，阴阳演变。将其作为灵感，打造不同风格的五个 VIP 接待室：金"皎月"、木"青竹"、水"清墨"、火"朱绯"、土"缃色"，五个主题风格相互对应，相辅相成。

皎月：以白色墙面、沙发、大理石地砖打造现代简洁空间，开阔大方，通体白色如皎洁月光。搭配暖黄色毛毯，在有些清冷的空间中融入一些温暖。落地窗引入自然日光，让房间内充满柔和而舒适的光线。

缃色：接待室将沙发设置成米色，并搭配明黄色小方椅、棕色休闲沙发，以偏欧式风格的优雅设计，让整个空间显得温馨明亮。黄色象征着欢悦、光辉和希望，能够让人身心放松。

朱绯：朱红色的沙发和墙面装饰画占据了空间的中心视线，配合同色系的窗帘、灯具、置物架，这个接待室给人最直观的感受便是"热情似火"，如同凤凰浴火重生。以红色为主色调，搭配黑色茶几、墨绿色沙发，让红色沉静下来，律动却不浮躁。

青竹：渐变绿色的窗帘、以植物为主要元素的绿色墙纸壁画、墨绿色的沙发以及半透明浅绿的茶几，让这个空间显得清新而有呼吸感，也生动诠释了绿色和"再生"的含义。

清墨：这个接待室以墨色为主色调，壁纸、沙发、主灯、茶几、窗帘，在白色空间中铺设大量的黑，打造沉稳空间。运用经典深色皮质、柔软皮毛材质和低调有气质的金属摆件、花束装饰，步入其中，安定而令人信任的气息立刻充斥感官。

Taking the "five elements" theory and the philosophy of Yin and Yang as inspiration, the designers created five VIP reception rooms with different styles: "bright moon" (gold), "green bamboo" (wood), "clear ink" (water), "red flame" (fire), and "pale yellow" (earth). The five themes/styles correspond and complement each other.

"Bright moon": White walls, sofas, and marble floor tiles are used to create a modern simple space, which is open and generous. The whole space is white as bright moonlight. Yellow carpets are used to add some warmth in the cool space. The French windows will introduce in natural sunlight to fill the room with soft and comfortable light.

"Pale yellow": The reception room features sofas in beige and brown and bright yellow small chairs, creating an elegant European style. The entire space feels warm and bright. Yellow is a symbol of joy, radiance, and hope, which can relax one's body and mind.

"Red flame": The Vermilion sofas and the painting on the wall are the focus of the space. With the curtains, lamps and shelves in the same tone, this reception room is about the "passion of fire", like a phoenix reborn. The main color is red, paired with a black coffee table and a dark green sofa, which help down the red color while maintaining its rhythm without being impulsive.

"Green bamboo": The gradient green curtains, green wallpaper with plant patterns, dark green sofas, and translucent light green coffee tables make this space appear fresh and breathable, vividly interpreting the meaning of green and "regeneration".

"Clear ink": This reception room is designed in ink color: wallpaper, sofas, main lights, coffee tables, and curtains. Black color is widely used in the white space to create contrast. Here you can find classic dark leather, soft fur materials, and understated and elegant metal decorations and bouquets. Once you step into it, you will instantly feel the calming atmosphere.

煜康荟美学生活馆
YUKANGHUI ART LIFE CLUB

2022 "永隆·星空间杯" 江苏省室内设计大赛金奖
2022 金堂奖餐饮空间杰出作品

■

业主单位：煜康荟文化发展有限公司
项目地点：江苏南京
面积：5000 平方米
完工时间：2022

Client: Yukanghui Culture Development Co., Ltd.
Project Location: Nanjing, Jiangsu Province
Area: 5,000 sqm
Completion Time: 2022

当我们谈论起东方文化，会产生怎样的联想？是故宫博物院里厚重的文物展品，历史人物写下的千万首诗词，还是雕梁画栋、飞檐翘角的古代建筑？当中国传统文化与当代设计碰撞、融合，XY+Z DESIGN 晰纹设计给出的答案是 "越东方，越世界"。

一切起因，源自南京。煜康荟位于南京玄武湖畔，背靠明城墙，临水而居，项目建筑为唐式风格合院。往前翻阅，南京是六朝古都、十朝都会，文脉传承而底蕴深厚；向未来去，南京则立足时尚前沿，跟随时代发展迅速。南京复合了经典与摩登，而在城林山河之上孕育而生的煜康荟也自带着东方与世界相融的基因。

When we talk about oriental culture, what do we think of? Cultural relics in the Palace Museum, thousands of poems written by historical figures, or ancient buildings with carved beams and painted cornices? When traditional Chinese culture and contemporary design collide and merge, the answer given by XY+Z DESIGN is: "The more oriental, the more international."

Yukanghui is located on the bank of Xuanwu Lake in Nanjing, with the Ming Dynasty city wall at its back and facing the water. The building is a courtyard house in Tang style. Nanjing used to be the capital during the Six Dynasties, with a rich cultural heritage. Now, Nanjing is standing on the forefront of fashion and developing rapidly. Nanjing is both classic and modern, and Yukanghui, nourished by this land, also has the gene of blending the East and the world.

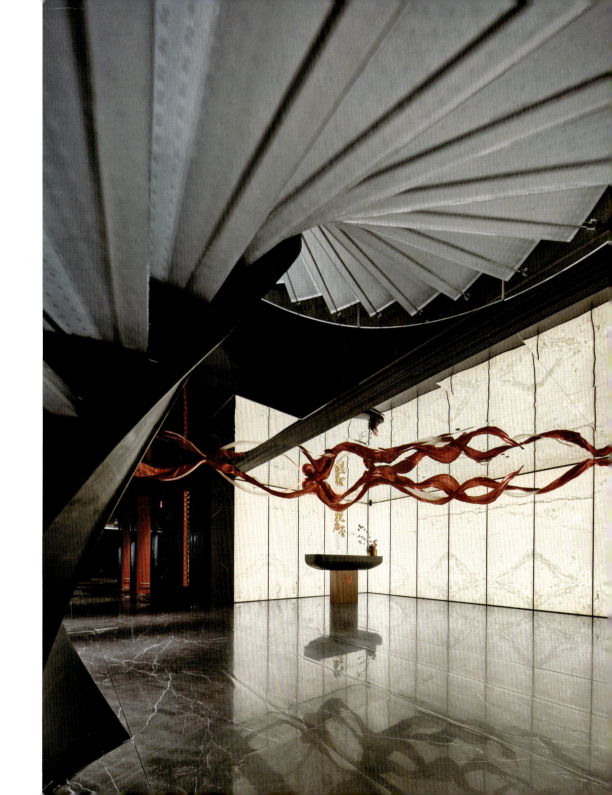

我们理解的东方、国风不单单是一味地延续古式，而是带着深刻的文化自信，去重新解读。以结合现代审美的艺术表现手法去构建空间，共话当代国风美学、时代趋势，同时融入低碳、环保的设计理念。踏入前厅，穿梭之旅即刻开始。悬浮于顶的飞舞绸缎装置结合镜面吊顶，通过光影反射，如朱砂在水中晕染。玻璃悬浮楼梯蜿蜒而上，直至煜康荟二层凤凰学宫。古今流转，时间与光影的灵动切换中，凤凰于飞，探寻匠心。

玉，非常完美地诠释及代表了东方文化之美。围绕在前厅的幕墙，采用天然玉石切片与玻璃组合的工艺手法。通过灯光穿透，好似山峰层叠的水墨画。降低耗材的同时，低碳、环保的设计理念得以彰显。

The Oriental and Chinese style, as we understand, are not just blindly continuing the ancient style, but reinterpreting it with profound cultural confidence. Here we tried to create a space by combining the ancient style and modern aesthetics, to communicate a new contemporary Chinese style, while integrating the design concepts of low-carbon lifestyles and environmental protection. As you step into the lobby, the shuttle tour begins immediately. The silk decoration is combined with the mirror ceiling, creating an interesting play of light and shadow. The suspended glass staircase winds up to the Phoenix Academy on the second floor.

Jade is used to represent and interpret the beauty of Oriental culture. The curtain wall around the lobby is clad with sliced natural jade and glass panels. When light penetrates, it looks like an ink painting of mountains. By reducing materials used, it coincides with the concepts of low-carbon lifestyles and environmental protection.

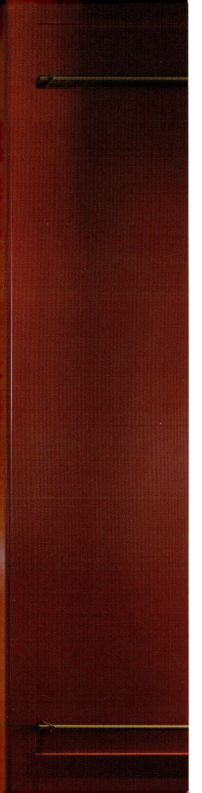

以"放""收""曲折""明暗""对称"等手法，表现空间的节奏与张力。用传统"闺阁"的"门"，作为"盖娅传说"的核心理念，通过朱红色对称的门扇，营造一个充满中国文化及当代仪式感的空间，为国风注入新的感悟。

在古人看来，屋顶不止有遮蔽天空的作用，它还是"技艺载道，道艺合一"的舞台。我们以"藻井"形式，摒弃传统藻井图案，用极简形式，呈现全新东方建筑形式之美，赋予展厅天顶之上。同时采用中轴对称的结构形式进行展厅平面布局，空间的张力由此为轴心展开。

The rhythm and tension of space are created by means of "opening", "closing", "twists and turns", "light and shade" and "symmetry". With the door of the traditional "boudoir" used as the core concept of "Heaven Gaia", a space full of Chinese culture and contemporary ritual feeling is created. The vermilion symmetrical door leaves inject fresh insights into the Chinese style.

In the view of the ancients, the roof not only acts as a shield of the house against the sky, but also serves as a stage for the integration of Tao and art. In a traditional but simplified form of caisson, we tried to present the beauty of new Oriental architecture by the roof. An axisymmetric layout is adopted, creating the tension of the space from the axis.

盖娅立体 Gaia Explosion

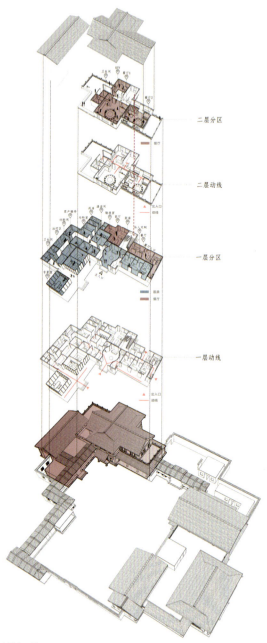

大健康立体 Health Center Explosion

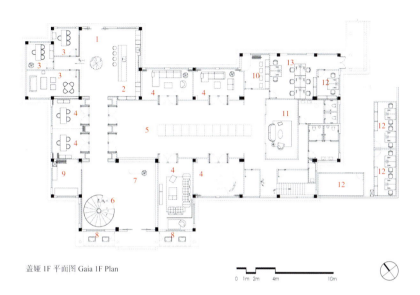

盖娅 1F 平面图 Gaia 1F Plan

0 1m 2m 4m 10m

1 吾道吾茶 Teahouse
2 烘焙区 Baking Area
3 茶艺区 Tea Art Area
4 贵宾接待 VIP Service
5 中岛台 Central Island
6 旋转楼梯 Spiral Stairs
7 前厅 Antechamber
8 橱窗 Display Window
9 打版间 Typesetting Room
10 化妆间 Dressing Room
11 多功能房 Multifunctional Room
12 办公室 Office
13 自由办公室 Free Office

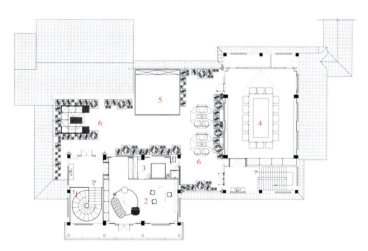

盖娅 2F 平面图 Gaia 2F Plan

1 旋转楼梯 Spiral Stairs
2 办公室 Office
3 卧室 Bedroom
4 会议室 Meeting Room
5 瑜伽室 Yoga Room
6 阳台花园 Balcony Garden

展厅内部在细节中传达设计品质，以传统东方美学构建出一个复合互动场景，"她"可以是展区、会议，或是一个秀场。一直以来，坚持"空间复合化设计"，营造丰富的场景变换，在空间中穿行，感受空间的艺术。

量身度体，唯主至上。"盖娅传说"高端定制服务是由内到外的一种艺术表达，六间风格各异的高端定制贵宾室秉持着"盖娅传说"的设计理念的同时，并融入"商圣六德"："静""智""勇""仁""强""礼"，共同构建了视觉与文明并存的空间之美。

Architectural quality can be found in details, creating an interactive scene with traditional oriental aesthetics. It can be an exhibition, a conference, or a show. The principle is "composite space design", creating rich scene variations.

"Heaven Gaia" is designed to provide high-end, customized service. Its six VIP rooms with different styles represent the "Six Virtues of Business": calm, wisdom, courage, benevolence, strength and courtesy. Modern visual interest and ancient civilization are combined into one space.

见最想见的人，喝最好喝的茶。吾道吾茶是我们的道场，我们的茶。

出于《书经》和《洪范》的五福之道—"长寿""富贵""康宁""好德""善终"，赋予五间贵宾茶室之灵魂。

人字顶、坡屋面，还原了中式建筑的形式，用最有"温度"的木制材料削减了现代设计手法中的硬朗，吐露温情。光的筛入透过竹帘细细绵绵，诗意翛然，清雅恬淡。

Meet the people you want to meet most and drink the best tea. That's what you will do in our teahouse.

The five VIP tearooms are named after the five ancient blessings in China: "longevity", "wealth", "health", "virtue" and "good ending".

The sloping roof is inspired by traditional Chinese architecture. One of the warmest materials-wood-is used to reduce the rigidness of modern design. Light penetrates in through the bamboo curtain, creating a poetic and elegant atmosphere.

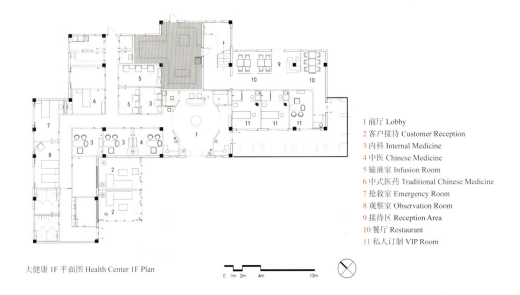

1 前厅 Lobby
2 客户接待 Customer Reception
3 内科 Internal Medicine
4 中医 Chinese Medicine
5 输液室 Infusion Room
6 中式医药 Traditional Chinese Medicine
7 抢救室 Emergency Room
8 观察室 Observation Room
9 接待区 Reception Area
10 餐厅 Restaurant
11 私人订制 VIP Room

大健康 1F 平面图 Health Center 1F Plan

0 1m 2m   4m   10m

以中式美学意境与现代审美艺术表现手法的结合去构造空间，是南京·煜康荟美学生活馆的基因。桐君还康健康管理中心将东方元素和现代时尚诉诸空间，通过极简的曲线流动于各处空间内，让生命的灵活展现于各处；曲面的结构，也让空间更具穿透与变化，使有限的格局中塑造丰富的视觉层次，实体的空间有了不设限的心境。

弧形吊灯自天花落下，暗金色的金属波纹，运用曲线螺旋结构与镜面天花于此巧妙邂逅。镜面设计的天花衍生了空间在视觉上的纵深感，让空间在交叠呼应的对话中，更加灵动丰盈。

It is the gene of the Club to create space by combining Chinese aesthetic conception with modern design techniques. In the Tongjun Huankang Health Management Center, oriental elements and modern design are combined. Minimalist curves flow through all the spaces. The curved surfaces also add connection and variation to the place, creating visual layers in the limited spaces.

The arc chandelier hangs from the ceiling. The spiral structure, like dark gold metal ripples, is reflected in the mirror ceiling, adding to the visual depth of the space.

所有美食的诞生，均以火炊之。橙红色岩板切片发光墙面与镜面交织，虚实相生，明暗相交，仿佛在述说时光的记忆。

The birth of all kinds of delicious food is based on fire. The light-emitting wall surfaces (orange red rock plates) are interwoven with mirror, where light and dark intersect, as if telling the story about memories of those joyous days...

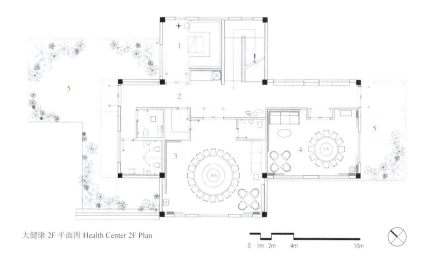

大健康 2F 平面图 Health Center 2F Plan

0 1m 2m   4m   10m

1 唱吧 Karaoke
2 廊道 Gallery
3 大包间 Large Private Room
4 小包间 Private Room
5 景观阳台 Landscaped Balcony

金陵文人辈出，以水而居，吟诗作画。

故而玄武湖畔的湖宴餐厅，以"水"为引，展开空间的设计。

宋代马远的"十二水图"，描述了世间所有的水之形态，画类无出其右。设计将十二水图作为软装主要元素与空间结合，表现水的灵动之美。

Jinling is the hometown of a large number of literati in ancient China.

They used to live by water, reciting poems and drawing paintings. Therefore, the Lake Banquet restaurant by Xuanwu Lake is designed with "water" as the inspiration.

Yuan Ma, a famous painter living in the Song Dynasty, depicted different forms of ripples in his painting "Twelve Waters". These are used to design the main elements of soft decoration to express the beauty of water.

包间顶部，竹简视觉形态的壁纸与现代却又复古的纸灯笼，共同构建中式人字顶，引人沉浸在空间之中，体会天地人和之乐。

In the private room, the wallpaper adopted on the ceiling that looks like bamboo slips and the modern but retro paper lanterns jointly build a Chinese gable roof, which makes people immerse in the space and experience the joy of harmony between heaven, earth and man.

宇宙奇迹生活艺术馆
XY+Z GALLERY

项目地点：中国上海
空间属性：商业空间
服务内容：EPC 总承包
建筑面积：600 平方米
设计时间：2022
完工时间：2022
主要材料：复合黑洞石玻璃、复合玉石玻璃、银镜、镜面不锈钢、
　　　　　　乳胶漆、地胶垫

Location: Shanghai, China
Program: Commercial space
Service: EPC (Engineering Procurement Construction)
Floor Area: 600 sqm
Design Time: 2022
Completion Time: 2022
Materials: Black travertine composite glass, jade composite glass,
　　　　　　silver mirror, mirror stainless steel, latex paint, floor mat

宇宙奇迹生活艺术馆作为综合性艺术空间，以"艺术设计新体验"为理念，由家具、艺术品、珠宝、咖啡、酒吧、展览活动等多种业态组成，反映生活品味的不同领域，鼓励人们探索空间、生活、美学、体验的新关系。从上海容纳百川的文化出发，融合场地天然优势，在空间设计上为人们带来更多感官沉浸式的复合化空间体验。

宇宙奇迹生活艺术馆位于上海中心城区东北部，地处黄浦江下游西北岸的杨浦区中心区域。其坐落的长阳创谷前身为中国纺织机厂，随着时代变迁对其进行了改造，改造并没有破坏原有的建筑结构，以原有的简约"工业风"为设计主要基调，把见证百年工业文化变迁的老厂房最大限度地保留了下来。

XY+Z Gallary, as a comprehensive art space, adheres to the concept of "new experiences in art design" and is composed of various elements such as furniture, artworks, jewelry, coffee, bar, and exhibition activities, involving different fields of life and encouraging people to explore new relationships between space, life, aesthetics and experience. Starting from the inclusive culture of Shanghai, based on the natural advantages of the site, we tried to bring more sensory and immersive experiences to visitors.

The Gallery is located in the northeast of downtown Shanghai, in the central area of Yangpu District on the northwest bank of the lower reaches of the Huangpu River. The building is the previous China Textile Machinery Factory. We renovated it without damaging the original building structure. The design is based on the existing simple "industrial style", and the old factory building, which has witnessed the century-old industrial and cultural changes, has been preserved to the greatest extent possible.

项目地处园区中心，本案设计将艺术融于生活，希望将设计师本身热爱的艺术品、珠宝、茶、酒、咖啡等汇聚于此。从外观、结构工艺、空间利用率等多个范畴考虑，将视觉感官、功能利用、品牌化在空间中更好地对外发散，打造一个集零售、策展、咖啡、酒吧等功能于一体的综合性艺术空间，在空间中每项业态的存在既是单一独立也是整体相叠，一同掌握空间的秩序。

出于园区场地条件、景观特征及周边商业形象考虑，由表及里，使业态、功能、形式得以有机连接。形成以"宇宙"为主题的整体设计，展现不同视角，延伸出设计的不同维度，构建起截然不同的设计逻辑。

The project is located in the center of the factory campus. The design tried to integrate art into daily life, hoping to bring together the designer's beloved artworks, jewelry, tea, wine, coffee, etc. here. Taking into account multiple aspects such as appearance, structure, and space utilization, we tried to combine visual effects, function and branding in the space to create a comprehensive art space that integrates retail, gallery, coffee, bar, and other functions. All the functions in the space are independent yet overlapping, jointly creating the order of the space.

Considering the site conditions, landscape characteristics, and surrounding commercial areas, we tried to organically integrate the businesses, functions, and their forms. Based on the theme of "universe", different dimensions of design are extended from different perspectives, completing completely different design logic.

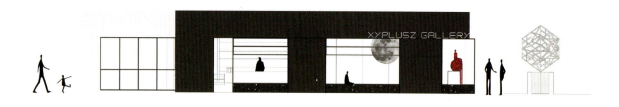

立面图 Elevations

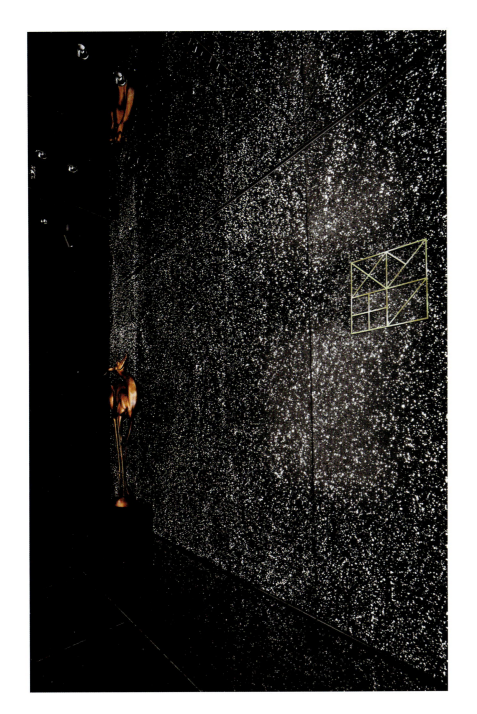

入口，成为探索宇宙之旅的起点

在门头的设计上，煞费心思。内外发光石板连成一片，营造出空间的层次关系，黑色则赋予整个空间神秘的氛围和格调。

前厅的空间是探索未知宇宙的序章，随着镜面天花与大理石地面的反射延伸出无限的空间，黑石、光点、神鹿，现代的艺术手法将其结合，用光影在如幻似虚的意境中体验空间美感。

The Entrance Becomes the Starting Point For Exploring the Universe

In the design of the storefront, great ingenuity is required. The inner and outer glowing stone slabs are connected to create a hierarchical relationship in the space, while black color gives the entire space a mysterious atmosphere and style.

The front hall is the prelude to exploring the unknown universe. With the reflection of the mirrored ceiling and marble floor, infinite space is extended. Black stone, light dots, and divine deer are combined with modern techniques, allowing light and shadow to play in the dreamy space.

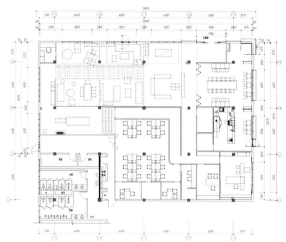

平面图 Plan

不止设计书展区

仿佛走进了设计师的个人宇宙，本案设计师也是本书作者十八年的设计生涯浓缩在书页铺成的穹宇，从对设计基础的理解到随后的实践拓展、设计上的个性表达以及一些具有代表性作品的再现。

流动的生命展区

从一滴水的独特视角，在有限的空间里，试图呈现自然界里生命的不同存在状态：经历极地冰川、热带丛林、荒野大地，最终回到生命的终点也是起点——海洋的怀抱。

"Beyond Design" Book Display Area

As if entering the designer's personal universe, the designer and the author of this book's 18-year design career is condensed in the vault of the page, from the understanding of the foundation of design to the subsequent expansion of practice, the expression of personality in design and the reproduction of some representative works.

Flowing Life Exhibition Area

From the unique perspective of a drop of water, in a limited space, the designer tried to present the different states of life in nature: from polar glaciers, to tropical forests, to the wild land, and finally returning to the endpoint (also the starting point) of life: the ocean.

「不止设计」书

设计师郭晰纹的新书《不止设计》
将于今年年底正式发行
本次展览作为该书的官方预展
首次向大众展示了书中的部分精彩内容

设计师18年的设计生涯
浓缩在书页铺成的穹宇
进入展区
仿佛走进了设计师的个人宇宙

从对设计基础的理解
到随后的实践拓展、设计上的个性表达
以及一些具有代表性作品的再现

不止设计的概念其实是想表达
所有的设计都来源于生活
就像本书作者对自己多重身份的洞察
继而引发出关于空间的复合性的思考
希望空间、材料、等不同存在形式的资源
能够发挥其最大价值是一切设计的初衷

· Glacier

The first thing that catches the eye upon entering the exhibition area is a "glacier" world built up with aluminum foil-a material commonly seen in daily life. This is the starting point of life. On this blue planet, the most insignificant yet incredible miracle is happening every moment-the flow of life.

· Jungle

The melting glaciers slowly flow towards the tropical jungle, and the miracle of life continues here. A miraculous tree, which took root with the company headquarters in Shanghai in 2015, once almost withered after months of trapped in 2022, due to lack of care. However, with persistent loving care from the founder of the company and his daughter, new leaves started to grow miraculously, showcasing the spirit of rebirth.

· 冰川

进入展览现场首先映入眼帘的是由生活中熟悉的铝箔纸装置构建而成的一片"冰川"世纪。这里是生命的起点，在这个蓝色的星球上，每时每刻都在发生着最微不足道也最不可思议的奇迹——生命的流动。

· 丛林

消融的冰川缓缓流向热带丛林，生命的奇迹在这里延续，这棵随公司总部一同于 2015 年在上海落地生根的奇迹之树，在 2022 年因受困数月，它曾一度因缺乏照料几乎被判为枯木，然而在公司创始人和女儿一段时间的悉心照料之下，它竟奇迹般地长出了新叶，焕发出更加耀眼的重生之美与感动。

· 大地 & 海洋

普通的干树皮和廉价的蓝色反光纸在设计师手中幻化出孕育生命奇迹的大地与海洋。自然生命的随机与有序、偶然与必然都蕴藏在眼前大大小小的树皮碎片里。所谓设计的奇迹往往就在于用最不起眼的材料制造最令人惊艳的感官体验。

· Earth & Ocean

Ordinary dry bark and cheap blue reflective paper have been transformed into the land and ocean that nurtures miracles of life. The randomness and orderliness of natural life, as well as haphazard and inevitability, are hidden in the bark fragments. The miracle of design often lies in creating the most stunning sensory experience with the least noticeable materials.

凝固的时间展区

时空之外，感官之间，一场凝固的美学盛宴正在上演。这是一场关于远古记忆的时间之旅，在此展出的每一款发光玉石，都是一件跨越时空的艺术作品。时间在这里被凝固，有了形状和模样，甚至可以用手轻轻去触摸，亿万年前的自然、生命、回忆和想象让感官在不同时间和空间里自由切换。

设计师的灵魂与时间一同凝固在如梦似幻的艺术空间中，浑身散发着华古而又优雅的光芒，将珍贵原料的隽永特质缓缓保存，通过手指的触摸亲自领略大自然的匠心，沉浸式的美学体验之旅也就自此开始。

平行时空长廊

原有的空间结构中，有着大量无法移动的立柱支撑整个建筑。而本案经过设计师合理精细划分空间，通往办公区域的时空走廊中全镜面的反射将把现实抽离，每个角度都关注到这个艺术走廊的不同光感映射形态和视觉互动，空间不再是以单维度存在。设计师寄望其能成为蕴含一切灵气的所在，不断地将"我"向四周渗透，立足于超越彼岸回望在此，为访客创造新的生命触动。

Frozen Time Exhibition Area

Beyond time and space, a feast of frozen time is unfolding before us. This is a time journey about ancient memories, and every piece of shining jade displayed here is a work of art that transcends time and space. Time is frozen here, with shapes and appearances that can even be lightly touched by hand. Nature, life, memories, and imagination billions of years ago allow the senses to freely switch between different times and spaces.

The soul of the designer and time are frozen together in a dreamy space. The space emits a magnificent and elegant light, showing the timeless characteristics of the precious materials. Through the touch of fingers, one can personally appreciate the craftsmanship of nature, and the journey of immersive experience begins from here on.

Parallel Spatiotemporal Corridor

In the original structure, there were a large number of immovable columns supporting the entire building. After the designer's reasonable and meticulous division of space, a "spatiotemporal corridor" with full mirror reflection leading to the office area is created. Here, space is no longer a single dimensional existence. The designer hopes that it can become a place that contains all spiritual energy, constantly infiltrating visitors into the surroundings, creating new life experiences.

宇宙会场

针对空间的"社交属性"，设计师倾心打造网红概念会议厅"宇宙会场"，黑色乳胶漆与镜面不锈钢成为氛围营造的最佳选择，大面积折叠窗将室内延伸到室外，折窗下可两侧对坐，内外浑然一体。银河流动，光点闪烁，充满工匠精神的质感与探索未知宇宙的神秘相辅相成。

咖啡厅

夜幕降临，设计师亲手绘制的月亮散发出迷人的光晕，一场沉浸式的星辰探秘，是心绪的流动与时间短暂凝固。在自然的空间韵律中，随着光影变幻传递出如同星球探索般的乐趣。

展示区作为本案主要空间，融合艺术品、软装、家具、珠宝、鞋履、纸艺、电器等多元化产品展示，以"优雅、时尚、前卫"的设计格调，塑造多元化个性且富有艺术灵魂的空间设计。

Cosmic Venue

Since it is a social place, the designer tried to create a conference hall called the "Cosmic Venue", which is intended to be a social media influencer. Black latex paint and mirrored stainless steel are the best choices for creating the desired atmosphere. Large folding windows extend the indoor space to the outdoor, and under the folding windows, you can sit on both sides, creating a seamless boundary between the interior and exterior. The "Milky Way" flows, with twinkling dots of light. The texture full of craftsmanship spirit complements the mystery of exploring the unknown universe.

Café

As night falls, the moon drawn by the designer emits a charming halo, creating an immersive experience where you can explore the stars. It is about the flow of emotions and the frozen time. In the natural rhythm of space, you can have the joy of exploring planets as light and shadow change.

The exhibition area, as a main space in the café, integrates diverse elements for display, such as art, soft furnishings, furniture, jewelry, footwear, paper art, and electrical appliances. With an elegant, fashionable, and avant-garde design style, a diverse and artistic space is created.

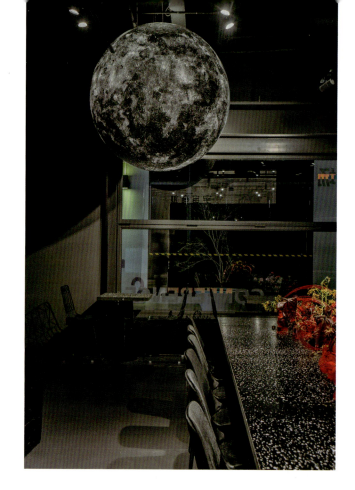

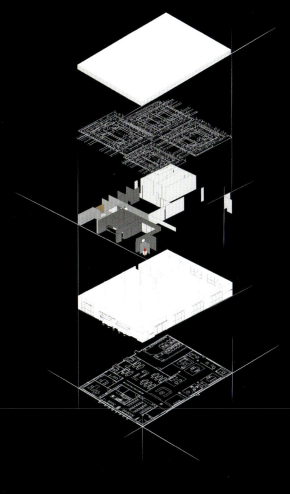

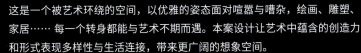

这是一个被艺术环绕的空间，以优雅的姿态面对喧嚣与嘈杂，绘画、雕塑、家居…… 每一个转身都能与艺术不期而遇。本案设计让艺术中蕴含的创造力和形式表现多样性与生活连接，带来更广阔的想象空间。

This is a space surrounded by art, facing the hustle and bustle of urban life with an elegant posture. Painting, sculpture, home... everywhere you will unexpectedly encounter art. The design tried to connect life with the creativity contained in art and the diversity of formal expressions, providing a broader space for imagination.

未来之光
LIGHT OF THE FUTURE

2017 "永隆·星空间杯" 江苏省室内设计大赛金奖
2020 她设计年度优秀空间设计奖
2020 中国室内设计大奖赛工程类优秀奖
2021 "永隆·星空间杯" 江苏省室内设计大赛银奖

■

业主单位：成都太行瑞宏房地产开发有限公司 - 成都·金沙城·锦
澜阁公区
项目地点：四川成都
面积：2214 平方米
完工时间：2019
设计材料：大理石、GRG、穿孔铝板、金属、艺术玻璃

Client: Chengdu Taihang Ruihong Real Estate Development Co., Ltd. /
Public Spaces in Kinlan Community, Jinsha Town, Chengdu
Project Location: Chengdu, Sichuan Province
Area: 2,214 sqm
Completion Time: 2019
Materials: Marble, GRG, perforated aluminum panels, metal, art glass

我们始终围绕着三位一体的未来人居生态理念：光、生态、未来。设计三角理念被运用于项目的整体规划并贯穿于各处细节，神奇的几何图案经由设计师之手成为一股来自未来的力量，牵引着整个空间的能量流动。

The design is always based on the ecological trinity of the future of living: light, ecology and future. The design triangle is used in the overall planning of the project and in every detail. The magical geometric pattern becomes a force from the future through the designer's hands, pulling the energy flow of the whole space.

我们对低区公共空间的功能进行合理、细致的规划布局，令其在与外界连接的同时，内部也构成一个功能相对完善而独立的生态系统。住宅部分的首层将原先的上人玻璃顶改造成玻璃水景，水的折射与反射因为光线变换而呈现更微妙的效果，从而既不影响自然光线又巧妙地让简单的物理效应阻隔了一二层商业区域对楼上住宅的干扰。

以每三层标准层作为一个微生态单元，在其首层设置自助水吧、室外花园作为该生态单元内的共享区域，为住户的日常生活提供便捷，提升品质。

The design provides a rational and detailed planning of the functions of the public spaces in the lower areas, so that while they are connected to the outside world, they also form a relatively well-functioning and independent ecosystem. The first floor of the residential section has been transformed from a glass roof to a glass water feature, where the refraction and reflection of the water takes on a more subtle form as the light changes, thus not affecting the natural light but subtly allowing the simple physical effect to block the commercial areas on the first and second floors from interfering with the residential areas above.

Each three standard floors are used as a micro-ecological unit, and a self-service water bar and outdoor garden are designed on the first floor of each unit, providing convenience and enhancing the quality of the residents' daily life.

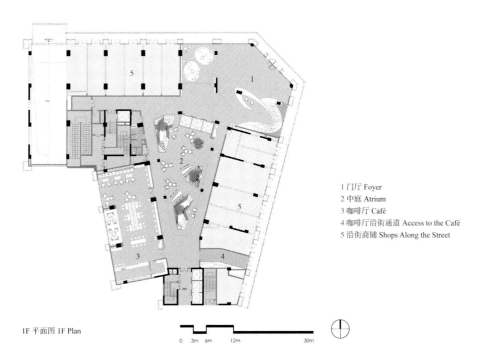

1 门厅 Foyer
2 中庭 Atrium
3 咖啡厅 Café
4 咖啡厅沿街通道 Access to the Café
5 沿街商铺 Shops Along the Street

1F 平面图 1F Plan

0 3m 6m   12m   30m

首层除了具有接待功能的大堂，轻餐饮与沿街零售空间的加入增强了其与周围社区的互动。二层规划以餐饮与健身功能为主，丰富住户的休闲娱乐生活，同时在地下一层，规划了洗衣房、自动贩卖机以及迷你仓等生活便利设施，令公共空间的功能更加完整。

标准层地面依然采用古木纹及白色瓷砖引导功能分区，白色金属板墙面让空间简净统一。入户门单独用黑色镜面不锈钢与金属木饰面包裹装饰，易于辨识。墙面主次关系分明，顶面的黑色镜面灯槽与白色乳胶漆，和地面古木纹与白色瓷砖呼应，侧面灯槽在合理包裹防火卷帘的同时发挥了装饰与视觉延伸的效果。

未来感的开放式设计体现了金沙城宜居、未来的氛围。极具科幻感的黑白灰色调，搭配灯光与流线的引导，表达了我们对未来居住环境的态度——简洁、舒适、健康、宜居。

In addition to the reception lobby on the ground floor, dining and street-level retail spaces are added to enhance the interaction with the surrounding community; on the second floor, dining and fitness functions are planned to enrich the leisure and entertainment life of residents, while on the basement floor, laundry, vending machines and mini-warehouses are planned to make the functions more complete.

The flooring of the standard floor is still made of antique wood grain and white tiles to guide the functional partitioning, while the white metal panel walls keep the space simple and unified. The entrance door is individually decorated with black mirrored stainless steel and metal wood veneer, making it easily identifiable. The walls are clearly defined, with black mirrored light sinks and white emulsion paint on the ceiling echoing the antique wood grain and white tiles on the floor, and side light sinks wrapped in fireproof roller blinds for a decorative and visual extension effect at the same time.

The futuristic open design reflects the liveable, futuristic atmosphere of the community. The futuristic black, white and gray tones, together with the lighting and streamline guidance, express our attitude towards the future living environment-simple, comfortable, healthy and liveable.

立面图 Elevation

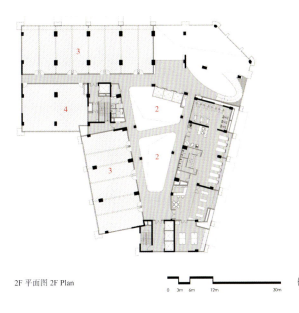

1 健身房 Gym
2 中庭上空 Void Above the Atrium
3 餐饮区域 Catering Area
4 厨房区域 Kitchen Area

2F 平面图 2F Plan

0   3m 6m   12m   30m

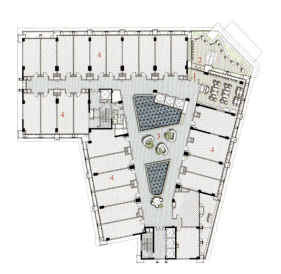

1 自助水吧 Self-Service Water Bar
2 室内休息平台 Indoor Seating Deck
3 中庭 Atrium
4 公寓 Apartment

3F 平面图 3F Plan

公区前厅接待区域地面采用 800mm×800mm 古木纹石材铺设而成，古木纹石材地面与走廊白色瓷砖形成鲜明对比，同时借由不同材料的纹理和色彩暗示相应的功能分区。地面材质自然延伸至楼梯，扶手曲线则呼应立柱造型与顶面曲线灯槽。树脂板与金属扶手的材质结合让楼梯造型更加灵活通透。

The flooring of the reception area is made of 800×800mm antique wood grain stone, which contrasts with the white tiles in the corridor, while the texture and color of the different materials suggest the functions of different zones. The flooring material naturally extends to the staircase, while the curve of the handrail echoes the shape of the column and the curved light groove on the ceiling. The combination of resin panels and metal handrails gives the staircase a more flexible and transparent shape.

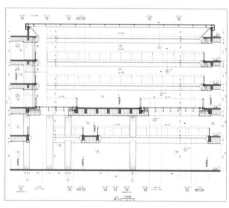

施工图 Construction Drawing

中庭的三角形室内花园作为项目的一大亮点，为建筑内部带来自然采光，搭配绿植丰富了内部的生态种类，使建筑中的人与自然产生多维度的链接。同时，中庭还增加了各楼层之间的可视性，进而促进了不同楼层住户之间的交流与互动。

透明玻璃轿厢观光电梯将带你从湖底到山顶多角度体验置身在这山水间的感受。让生活在这里的每个人关上门拥有一个家，打开门拥有一座山。

过道区域地面运用黑色石材与白色瓷砖将休息区与走道区自然分割。墙体采用金属板装饰，暗藏灯带的设计给予空间动线引导，楼板侧面亦是如此，并与顶面不规律的三角形元素金属板一起，共同呼应了设计的主题。

The triangular-shaped indoor garden atrium is a highlight of the project, bringing natural light to the interior of the building and enriching the ecological variety of the building with greenery, creating a multi-dimensional link between people and nature within the building. The atrium also increases the visibility between floors, thus increasing the communication and interaction between residents on different floors.

The transparent glass lift will take you from the bottom of the lake to the top of the hill to experience the feeling of being in the landscape from multiple angles. It allows everyone living here to have a home behind closed doors and a mountain behind open doors.

The aisle area is naturally divided between the lounge area and the walkway area by the use of black stone and white tiles on the floor. The walls are decorated with metal panels with concealed light strips to guide the movement of the space, the sides of the floor are also decorated with metal panels with concealed light strips to echo the overall elements, while the metal panels on the ceiling use irregular triangular elements to echo the theme of the design.

电梯厅墙面延续黑白灰的主题色，黑色镜面不锈钢包裹电梯门与收口，电梯与电梯之间做出一个个玻璃盒子，四周设置灯光，内部图案呼应前厅及过道区的设计元素。

整段旅程糅合了诸多的对立元素：围合与开放、明与暗、精细与厚重……从空间的营造到灯光的处理，每一个细节都体现了都会生活进阶的全新诠释——让访者在置身于此的每时每刻都能够感受到神秘和惊喜，激发探索的欲望，怀着热烈和愉悦的心情感受每一寸空间和每一件艺术品。

The walls of the lift lobby continue the theme of black, white and gray, with black mirrored stainless steel wrapped around the lift doors and closures, and a glass box placed between each two lifts, with lighting around it and internal patterns echoing the design elements of the vestibule and aisle area.

The whole journey combines many opposing elements: enclosure and openness, light and darkness, refinement and heaviness... From the creation of the space to the treatment of lighting, every detail reflects a new interpretation of urban living progression - allowing visitors to feel mystery and surprise at every moment of their stay, inspiring the desire to explore and to experience every inch of space and every work of art with enthusiasm and pleasure.

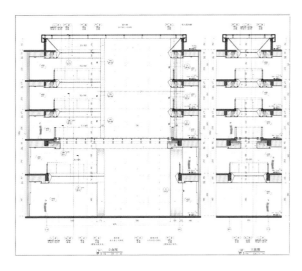

施工图 Construction Drawing

时间换空间
TIME FOR SPACE
一尺山房｜家中博物馆
HILL VILLA

这里应该算是我室内设计的一个里程碑。

"时间换空间"的甲方领导范总，也是"不纸书店""金沙书画院"的甲方。他对我的设计有很大影响，无论理念还是方法。印象最深的是"矛和盾不一定对立，也可以平衡"，以及如何做到"叫好又叫座"。名副其实的"良师益友"。在此，正式地对他鞠躬感谢。"时间换空间"，这个名字是他的定义。人们对空间，在不同的时间有着不同的需求。而一个普通小小酒店式公寓的业主，常常苦恼的就是空间太小，不够用。那么怎么让"小"不再是"小"，就是这个项目的主要目的。通过两年的研发设计、打样过程，该项目的室内装配率达到了95%。也完成了真正"时间换空间"的使命，更集智能化于一体。在此，除了感谢甲方外，还要感谢我的同事们的配合及付出，感谢梅亚奇工厂对执行及技术方面给予的支持及帮助。

"一尺山房"，与本书第二大类"展示与智能空间的融合"内多个项目的业主为同一人。由这些不同空间所呈现的完全不同的风格可以看出我对居住空间的理解及认知。一个空间，究竟是风格及美更重要，还是让居于其中的人舒适、便于使用更重要？也就是说，居住空间设计究竟是给人看的还是给人用的？杨院长说，

居住的空间理解

UNDERSTANDING OF LIVING SPACE

广义上的居住空间设计，应该是一种交互式的行为艺术。从这个意义上说，设计的交付只是工程施工部分的结束，并不是设计的完成。后一部分将由业主更多参与，把人和环境的因素及理解植入，共同互动完成。居住空间设计的核心在强调时间的表现，更多呈现时间变化的空间设计是有生命的设计。这是最难也是最复杂的。因为，生活和烟火气息息相关，空间与生活的时间无法分割。想象一下，如果你家白得、干净得像个艺术馆，这种干净，能持续多久？家，是每个人心里最重要的地方，随着时间的推移，会有不同的"伙伴""记忆""故事"融入这个空间。所以，必须能承载居住者的爱好、习惯并且伴随其一路成长。人、物、习惯、气味、声音、光线的变化都将在此结合。空间会变成一个个有内涵的场景，而这样的场景，须要有人的参与才鲜活，才叙述。随着时间流逝，会不断有新的内容增加和空间的调整，都会带来新的变化。空间在时间的滋养下不断生长，变幻出不可思议的结局，这是设计的最高境界，也是设计的生命所在。

This could be considered a milestone in my interior design career.

The client, Mr. Fan, who believed in the concept of "time for space", was also the client for Buzhi Bookstore Chengdu and Jinsha Art Center. His influence on my design was significant, both in terms of concept and method. The most memorable for me was his idea that "spears and shields are not necessarily opposing forces; they can also balance each other out" and "you have to find a way to achieve both popularity and commercial feasibility". He was truly a "mentor and friend". I would like to formally express my gratitude to him. "Time for space" was his definition. People have different needs for space at different times. The owner of an ordinary small hotel-style apartment often has the problem of insufficient space. So, how to make a small space not small anymore was the main goal of this project. After two years of research and design, the interior assembly rate for this project reached 95%, achieving the true mission of "time for space" and integrating intelligence into it. Here, I would also like to thank my colleagues for their cooperation and dedication, and thank the Meyach factory for its support and assistance in implementation and technology.

Hill Villa has the same owner as some of the projects in the second part of this book, "integration of display and intelligent space". From the completely different styles in these spaces, you can see my understanding and cognition of living space. Is showcasing style and elegance more important for a space? Or is it more important to make the people living in it comfortable and easy to use? That is to say, is the design of living space for people to see or for people to use? President Yang said that the design of living space in a broad sense should be an interactive performance art. In this sense, the delivery of the design is only the end of the construction part, not the completion of the design. For the latter part, the owners will be involved, embedding the factors of people and space, which will interact with each other. The core of residential space design emphasizes the performance of time, and the space design that shows more time changes is a living design. This is the most difficult and complicated. Because life is closely related to the vulgar world, and space and the time of life cannot be separated. Imagine if your home is white and clean like an art gallery, how long will this cleanness last? Home is the most important place in everyone's heart. As time goes by, different "partners", "memories" and "stories" will be integrated into this space. Therefore, it must be able to accommodate the hobbies and habits of the occupants and grow along the way. Changes in people, things, habits, smell, sound, and light will all be combined here. Spaces will become connotative scenes, and such scenes require the participation of people before they can be vivid and narrated. With the passage of time, there will be new additions and adjustments, which will bring new changes. Space continues to grow with the nourishment of time, and you will have incredible endings. This is the highest level of design and the soul of design.

时间换空间
TIME FOR SPACE

2017 "永隆·星空间杯" 江苏省室内设计大赛银奖

■

业主单位：成都太行端宏房地产开发有限公司 - 成都金沙城未来公
寓研发设计
施工单位：四川中辰城建装饰设计工程有限公司
家具执行：南京梅亚奇家居用品有限公司
智能化执行：上海雷盎云智能技术有限公司
项目地点：成都太行端宏金沙城
面积：35 平方米
完工时间：2019

Client: Chengdu Taihang Duanhong Real Estate Development Co. /
Future Apartments, Jinsha Town, Chengdu
Constractor: Sichuan Zhongchen Chengjian Decoration Design &
Engineering Co.
Furniture Execution: Nanjing Meiyaqi Home Furnishing Co.
Intelligent Execution: Shanghai Lei Ang Yun Intelligent
Technology Co.
Project Location: Chengdu, Sichuan Province
Area: 35 sqm
Completion Time: 2019

在这个地价攀升、房价上涨的时代，如何用最小的空间满足最大的需求，是我们一直在探索的话题。每个人在生活中都扮演着多种角色，拥有着复合化的身份。我们相信，在不同的时间里，空间也可以被赋予不同的功能及使命。为满足当代人对物质及精神的双重追求，设计师利用时间的不同纬度，换取不同的空间功能。由此，"时间换空间"这个探索性设计产品应运而生。

In this era of escalating land prices and rising house prices, the question of how to use the smallest space to meet the largest needs is a topic we have been exploring. Each person plays multiple roles in life and has a composite identity. We believe that at different times, space can be given different functions and missions. In order to meet the dual pursuit of material and spirituality of contemporary people, the designers use different latitudes of time to exchange for different spatial functions. The result is Time for Space, an exploratory design product.

这是一个全国首家全智能多功能转换的空间，是设计团队自主研发的施工、家居、产品一体化设计空间。设计师研究人在不同时间段的空间行为，创意性地将智能化科技手段，植入这个集人性化、强大储藏性等特点，仅 35 平方米的高品质酒店式公寓中。

This is the first fully intelligent multi-functional conversion space in China, which is the design team's independent research and development including design, construction, and products. The designers considered the spatial behavior of people in different time periods and creatively used intelligent technology in this high-quality serviced apartment in only 35 square meters.

项目亮点：研发智能移门电视、未来厨房，以及区域智能切换功能，均已申请专利保护。

现代简约设计：简洁明快，重视室内美观效果及功能实用的结合，新型环保、科技材料的应用，使室内空间更通透、安全性更高、实用性更强。

智能化：四种灯光智能控制场景，自由切换满足灯光氛围营造。四个智能控制区，以智能移门电视为核心控制区，可控制大部分电动家具产品及场景切换。家电通过触摸、升降、感应等多种方式，给人们带来快捷智能科技体验，如黑科技穿衣镜、未来厨房等。

人性化、实用性：家具尺度方便灵活，多功能写字台能转换成梳妆台、飘窗榻；自由切换的电动升降床；灵活组合的魔术书柜，其隔板能自由切换。以上种种，共同构成了多功能的生活状态。

复合场景：会客、书房、单卧室、双卧室、餐厅及节能状态等多种模式。

Project highlights: development of intelligent sliding door TVs, future kitchens, and intelligent switching of zones, all of which have applied for patent protection.

Modern minimalist design: simple and bright, emphasizing the combination of aesthetic effect and functional practicality of the interior, the application of new environmentally friendly, technological materials, making the interior space more permeable, safer and more practical.

Intelligent: four lighting intelligent control scenes, freely switchable to create different atmospheres. Four intelligent control zones, with the intelligent sliding door TV as the core control zone, can control most electric furniture products and scene switching. Home appliances bring people a quick intelligent technology experience through touch, lift and induction, such as "black technology" dressing mirror and future kitchen.

Humanization and practicality: the scale of furniture is convenient and flexible, with multifunctional writing desks that can be converted into dressing tables and floating window couches; electric lift beds that can be switched freely; and magic bookcases with flexible combinations whose partitions can be switched freely. All of the above together form a multifunctional living state.

Compound scenes: meeting rooms, study rooms, single bedrooms, double bedrooms, dining rooms and energy-saving states, and many other modes.

线稿图 Sketch

STANDARD MODEL:
ONE BEDROOM, ONE LIVING ROOM AND ONE BATHROOM

进入一室一厅一厨一卫的空间功能模式，全方位满足客厅、卧室、厨房、洗手间的刚性需求。

The standard model (one bedroom, one living room, one kitchen, and one bathroom) will provide the basic functional requirements.

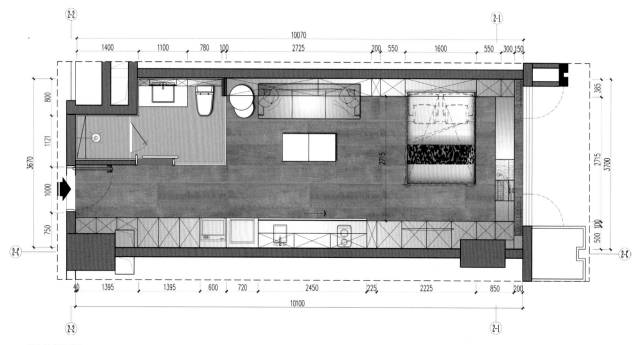

标准模式平面图 Standard Model Floor Plan

DINING MODEL: ONE BEDROOM, TWO LIVING ROOMS,
ONE KITCHEN AND ONE BATHROOM

就餐时间，一键开启就餐模式，进入一室二厅一厨一卫的功能模式，娱乐与用餐并用的复合空间，厨房展露，餐桌升起，美味在此开始。

It's dining time! Press a button and you will start the dining mode: one bedroom, two living rooms, one kitchen and one bathroom. Entertainment and dining are combined in one space. When the kitchen opens and the dining table rises, enjoy a delicious meal here!

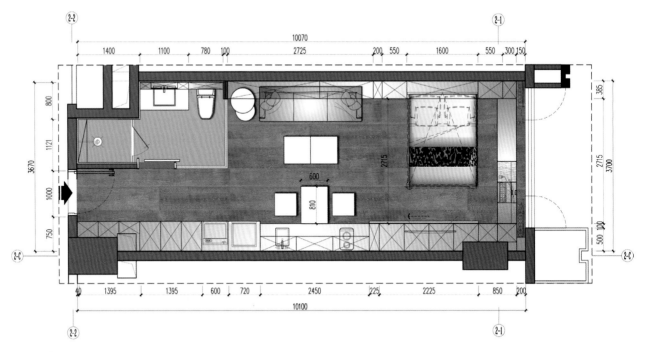

就餐模式平面图 Dining Model Floor Plan

双卧 模式 | 两室一卫

DOUBLE BEDROOM MODEL:
TWO BEDROOMS AND ONE BATHROOM

亲朋留宿，通过转换开启双卧模式，多增加一个卧室，满足多人对睡眠空间的需求，空间清爽、有序，富有现代感和整体美。

When families and friends stay overnight, you can start the double bedroom mode. One more bedroom will be created to accommodate more people sleeping. The simple, orderly layout adds to the quality modern design.

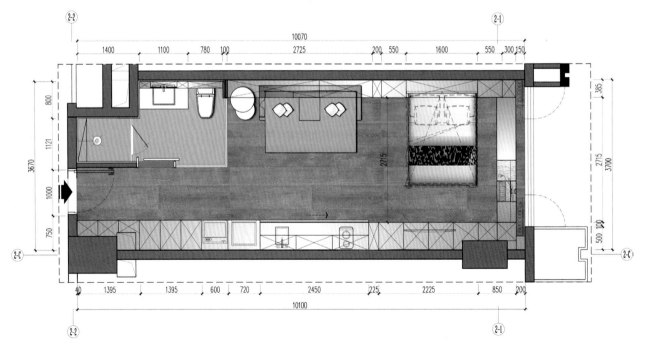

双卧模式平面图 Double Bedroom Model Floor Plan

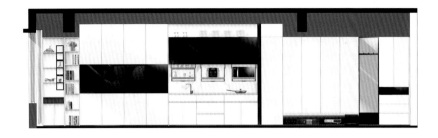

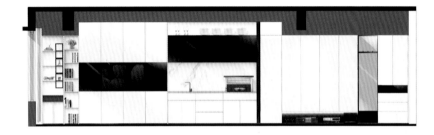

场景立面图 Elevations

未来厨房：2.4 平方米的平面空间，全智能化厨电，包含了厨房的所有功能，像变形金刚一样。电动升降吊柜隐藏着电饭锅与微波炉等家用电器；隐藏式感应升降油烟机、电动碗碟篮、抽拉餐台、米箱面箱、智能水槽洗碗机、净水处理器及垃圾处理器等共同构成这个极具未来感的智能厨房。

The kitchen of the future: The 2.4 sqm space is equipped with all kinds of intelligent kitchen appliances, containing all the functions a kitchen needs. The hanging cabinet hides the rice cooker and microwave oven and other household appliances. The hidden range hood that can be lifted or lowered through the sensor, electric dish basket, pull-out dining table, rice box and flour box, intelligent dishwasher, water purifier and garbage disposer together complete this very futuristic and intelligent kitchen.

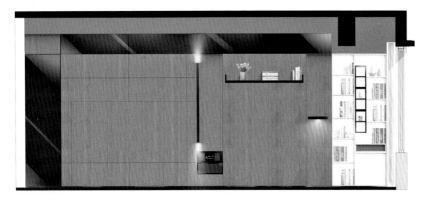

书房与梳妆台转换：黑色的格子通过模块化形式，可以自由地移动调到不同位置以满足人们不同的需求。多功能写字台可转换为梳妆台、飘窗榻等不同功能；飘窗台下为餐厅凳收纳处，同时餐厅凳兼具收纳箱、移动座位等功能。

Study or dressing table: The black grid can be freely changed to meet different needs through the modular system. The multi-functional writing desk can be converted into a dressing table or a window couch. Under the window couch is room for stool storage, while the dining stools have functions such as storage and mobile seats.

场景立面图 Elevations

一尺山房 | 家中博物馆

HILL VILLA

设计师的别具匠心在于：强调时间的表现，更多呈现空间中时间的变化，做一个有生命的设计。这是最难也是最复杂的，它重新定义了：设计的完成状态；空间对时间的表达；时间的静态与动态表达；设计中软装与硬装的关系。

项目位于江苏宝华山，此山因南朝梁代高僧登山结草为庵，设坛讲经传教而闻名。因其深厚的文化积淀，故在此山间别墅的设计中，设计师不仅运用了"借用"光线和风景等手法，更试图探索对于空间设计中"时间"的表达。

■

项目地点：江苏省宝华山大山地
原建筑面积：390 平方米
改造后建筑面积：560 平方米
完工时间：2019

Location: Baohua Mountain, Jiangsu Province
Existing Floor Area: 390 sqm
Floor Area After Renovation: 560 sqm
Completion Time: 2019

The designer's wisdom lies in emphasizing the expression of time by presenting the changes of time in space, creating a "living design". This is the most difficult and complex task, trying to redefine: the completion state of the design; the expression of time in space; the static and dynamic expression of time; the relationship between soft and hard decoration in design.

The project is located in Baohua Mountain, Jiangsu Province. This mountain is famous for the temple built by monks in the Southern Liang Dynasty. Due to this profound cultural heritage, in the design of this mountain villa, the designer not only utilized techniques such as "borrowing" light and scenery, but also attempted to explore the expression of "time" in space.

诗云"一尺过江山",说的不仅是小中见大,细微里见恢宏,更是如何在尺幅方寸之间,见天地、见历史,见空间、见时间。东西千万里,上下五千年。项目命名"一尺山房",正是为了尝试以"一尺"写"江山",以当下见古今的设计目的。

The Chinese name of the project comes from an ancient poem. It described the wisdom of seeing much in little, a way to view heaven and earth, history, space, and time. Thus, the goal of design is to reflect the past in the present.

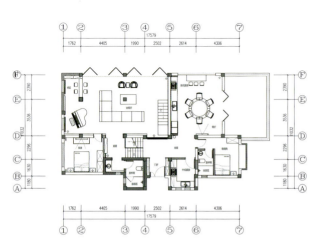

1F 平面图 1F Plan

2F 平面图 2F Plan

本案的业主是科学家亦是艺术家，同样还是收藏家和艺术史研究者。设计师从业主的身份职业背景、个人兴趣爱好等多个层面出发，重点体现天-地、山-水、人-物、虚-实之间的关系，人与建筑、环境的关系，以此诠释业主的气质与内涵。

设计师与业主充分沟通，认同只有把光阴和光同时纳入空间，才是设计的完成。从这个意义上说，设计的交付只是工程施工部分的结束，并不是设计的完成。后一部分将由业主更多参与互动完成，需要业主把人和环境的交流与理解植入空间。因此，广义上的设计应该是一种交互式的行为艺术。

The owner of the villa is a scientist and artist, as well as a collector and researcher in art history. The designer focused on reflecting the relationships between heaven and earth, mountains and water, people and objects, and virtuality and reality, as well as the relationships between people, architecture, and the environment, from multiple aspects such as the owner's identity, professional background, and personal interests and hobbies, in order to interpret the owner's temperament and style.

The designer and the owner fully communicated and agreed that only by incorporating both time and light into the space can the design be completed. In this sense, the delivery of design is only the end of the construction part, not the completion of the design. The latter part will be completed with participation and interaction of the owner, who need to implant communication between people and the environment into the space. Therefore, in a broad sense, design should be an interactive form of performance art.

空间中时间的表达往往被忽略，也是设计上最复杂的部分。设计师认为空间设计的最高境界在于对时间的理解和表达，可将影视戏剧中的舞美与叙事手段借鉴融入，形成独有的时空交错的设计美学。

The expression of time in space is often overlooked and is also the most complex part of design. The designer believes that the highest level of interior design lies in understanding and expressing time, which can be inspired by the stage art and narrative techniques in film and television dramas, forming a unique design aesthetic where time and space interweave.

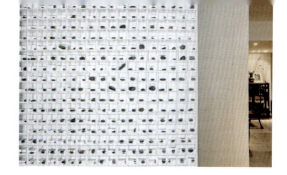

这其中包括光阴流淌——时间的静态与动态表达；情绪设计——软装与硬装的关系；沉浸体验——声音与画面的融合；快与慢的节奏——距离与简繁的搭配。

在业主数十年于世界各地的游历中，收藏了各种有"故事"的艺术品，这些艺术品成为本案首要的构成元素。设计师从这些记录了光阴变化、世事变迁的元素里，将不同时期的文化、主题进行排列，随着不同风格情调的音乐、檀香和光线变化。欧洲的铜版画、古典低腰钢琴、数百年历史的旧版书、中国的古籍善本、传统的中式家具、欧洲橡木与胡桃木的老式家具……在空间中彼此对立融合，相映成趣。

This includes: the flow of time (the static and dynamic expression of time); emotional design (the relationship between soft and hard decoration); immersive experience (the fusion of sound and visuals); fast and slow rhythm (involving distance and complexity).

During decades of travels around the world, the owner of the villa collected various artworks, which have become the primary elements in the design. The designer tried to arrange these elements according to their themes, periods and cultures, matched up with different styles of music, sandalwood, and light. European copperplate prints, classical low rise pianos, old editions of books with hundreds of years of history, rare ancient Chinese books, traditional Chinese furniture, old-fashioned furniture made of European oak and walnut... all blend in opposition and complement each other in the space.

"一尺山房"，不能用某一种特定的风格去形容，设计只是在表达"家"的模样。温暖、舒适、阳光、自由、轻松……处处都是一个生活的片段。

对时间的设计能否借鉴蒙太奇手法来处理？

设计综合运用音乐、香味、影像、收藏品等元素，与光线的变化相结合，创造出令人目眩神迷的空间单元："压缩了的时间""错乱了的时间""延长了的时间""被尊重的时间""遗失了的时间"等。空间变成了一个个有内涵的场景，而这样的场景，须有人的参与才鲜活，才能叙述。随着时间流逝，业主新的展品又会加入空间，这里又会有新的变化。

Hill Villa cannot be described with a specific style. The design aims to show the appearance of "home". Warm, comfortable, sunny, free, relaxed... Everywhere you would find a fragment of life.

Can we use montage techniques to design the expression of time?

The design integrates elements such as music, fragrance, images, and collectibles, combined with changes in light, to create some dazzling and mesmerizing "spatial units": "compressed time", "disordered time", "extended time", "respected time", "lost time", etc. Space has become a series of meaningful scenes, and such scenes require the participation of people to be vivid and narrated. As time passes, new exhibits from the owner will be added to the space, and there will be new changes here.

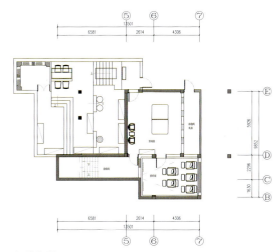

地下一层平面图 B1 Plan

如阁楼书房的地板，是专赴苏州挑选的老地板，通过与切割成同等尺寸的玻璃交错排列，使阳光投在楼道内时呈现出业主熟悉的儿时印象。拾级而上，去摩挲不同年代的木刻、铜板、石雕、银铸的文化沉淀，都让人生出无限遐想，常常忘记时间。此刻情绪的交互与音乐的感染，形成了思想的空间、情感的空间。

空间在时间的滋养下不断生长，变幻出不可思议的结局，这是设计的最高境界，也是设计的生命所在。

设计始终没有终结。

"一尺山房"将以它自己的方式，不断生长，自然成长。

For example, the floor in the study in the attic is paved with recycled materials specially sourced from Suzhou, arranged in a staggered manner with glass cut to the same size, creating a familiar childhood impression for the owner when sunlight shines into the corridor. Stepping up via the stairs, you can touch the woodcuts, copperplates, stone carvings, and cast silverware made in different eras, which will evoke infinite imagination, and make you forget time. At this moment, the interaction of moods and music will form a space for thoughts and emotions.

With the nourishment of time, the space will continue to grow and transform into incredible endings, which is the highest level of design and the life of design.

The design has never come to an end.

Hill Villa will continue to grow and develop naturally in its own way.

本案积累一些设计过程体会和经验与大家共同分享：

·设计的留白与留空：墙的处理，浮雕、镂雕与透雕的咬合，空间是看不见的道具和材料。

·设计的肌理与再生：硅藻泥、木质、石材和布艺的分时搭配。

·设计的音效与视觉：不同场景单元里声场、色彩和情绪的识别。

·设计的节奏与风格：东西方元素、简约与繁复、动与静、距离和转折的把握。

·设计的透视与场景：散点透视与焦点透视、移步换景和悬念的创建。

·设计的虚实与起止：不设计的设计、虚拟化的设计和设计师留给业主的设计，以及设计的生命力所在——自我生长与变化、迭代空间。

I'd like to share with everyone the following experiences and insights in the design process:

· Blank space or empty space in design: The treatment of walls, the interlocking arrangement of reliefs and carvings space is an invisible material.

· The texture and regeneration of design: Diatomaceous mud, wood, stone, and fabrics are combined.

· Sound and visual effects: The different scene units can be easily identified by sounds, colors, and emotions.

· Rhythm and style: Eastern and Western elements, simplicity and complexity, dynamic and static state, distance, and contrast are used.

· Perspective and scene: Different techniques are used, including cavalier perspective and focal perspective, "changing scenery", and suspense.

· Virtuality & reality, beginning & end of design: It's a design without designing, a virtualized design, and a design left by the designer to the owner. Here the vitality of design lies in self-growth and change of space – an iterative space.

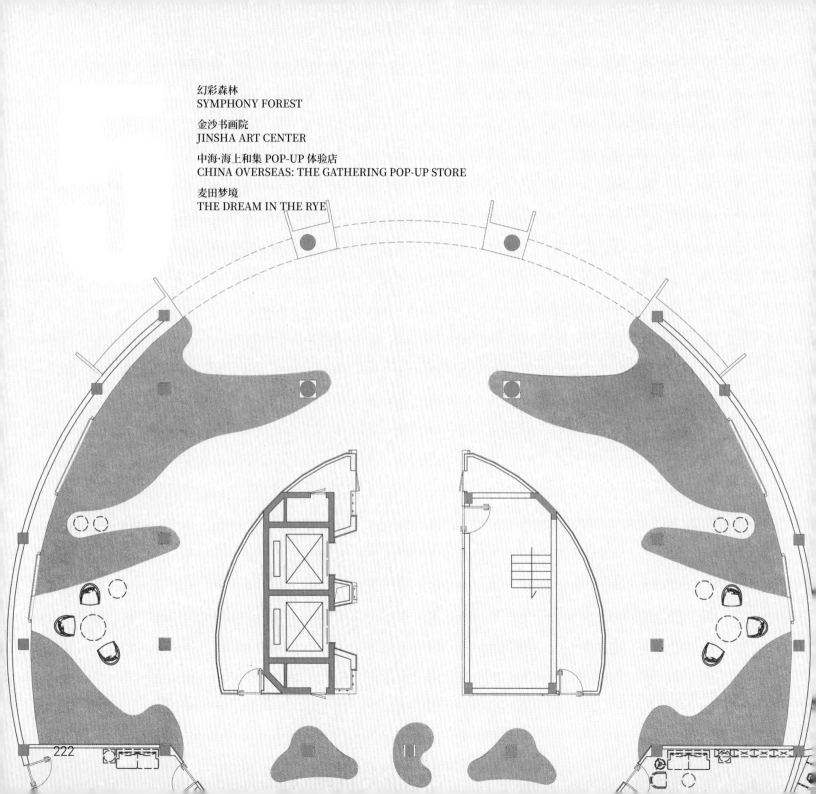

幻彩森林
SYMPHONY FOREST

金沙书画院
JINSHA ART CENTER

中海·海上和集 POP-UP 体验店
CHINA OVERSEAS: THE GATHERING POP-UP STORE

麦田梦境
THE DREAM IN THE RYE

222

城市更新及建筑与室内一体化设计案例

URBAN RENEWAL, ARCHITECTURAL AND INTERIOR INTEGRATION DESIGN CASES

我的先生——吴耀隆，多年来一路陪伴，给予我太多启发以及帮助。先生毕业于同济建筑专业，有丰富的设计院、地产甲方设计管理等建筑设计经验。我们因 2012 年 "同济米兰设计管理硕士班" 相识。建筑师、有品位、有思想、身高一米九五……最重要——长得帅。我必须承认自己在这方面的肤浅——喜欢一切好看的。从此，因为我俩的结合，建筑与室内，真正的融合。

大多数时候，室内和建筑很难有沟通配合的机会，因为大家都习惯 "这是建筑的事" "这是幕墙的事" "这是室内的事" "这是软装的事" "这是机电、暖通的事"……种种推诿，我也理解，准确来说，大部分确实不算自己的事。最后就是，太多的无法衔接或者格格不入。我认同项目设计负责人制，他应该统筹整体的设计。当然，我的这种认知和理解，也时常导致我和我的团队会很累。因为我会主动去对接其他专业，甚至给出解决方案，但往往别人会觉得我 "多事"。如果遇到一体化设计的项目，其实我会特别开心。因为我不再是 "多事"。下面列举几个项目，说一说一体化设计的重要性。当然，在日新月异的当下，对于 "城市更新" 这个名词，我们已经见得太多。过去，是存在过的，我认为不应该抹灭它。而现在，是越来越好的当下。如同生命一般，不能倒退。对于老旧建筑，我们应当在敬畏的同时，赋予它全新的生命。

My husband, Yaolong Wu, has accompanied me for many years, providing me with inspiration and help. He graduated from Tongji University with a major in architecture and has extensive experience in architectural design management for design institutes and real estate developers. We met in 2012 in the "Tongji Milan Design Management Master Class". An architect with taste, ideas, 1.95 meters tall... and most importantly, handsome. I must admit my shallowness here. I love everything that looks good. Since our combination, architecture and interior design have truly merged.

Most of the time, it is difficult for interiors and architecture to communicate and cooperate because everyone is used to saying, "this is an architectural matter," "this is a curtain wall matter," "this is an interior matter," "this is a soft decoration matter," "this is a mechanical and electrical, HVAC matter." Countless excuses. I understand, accurately speaking, most of these are not their own issues. In the end, there are too many things that cannot be connected or incompatible. I agree with the project design leader's responsibility for overall design coordination. Of course, my understanding often leads to fatigue for me and my team, because I will take the initiative to coordinate with other professions and even give solutions. But often, others may think I am "meddling". If I encounter a project with integrated design, I will be particularly happy because I am no longer "meddling". This part is a collection of several projects to illustrate the importance of integrated design. Of course, in the rapidly changing present, we have seen too much of the term "urban renewal". In the past, it existed and I don't think we should erase it. Now, it is getting better and better, just like life. We cannot go back. For old buildings, we should give them new life while respecting them.

幻彩森林
SYMPHONY FOREST

■

业主单位：桐乡市濮院毛衫发展股份有限公司 - 濮院时尚工场
项目策划：上海设界商务信息咨询有限公司 陆平一、杨志强
项目地点：浙江桐乡
面积：23000 平方米
完工时间：2020

Client: Tongxiang Puyuan Sweater Development Co., Ltd. /Puyuan
        Fashion Park
Project Planning: Shejie Business Consulting Co., Ltd., Shanghai /
        Pingyi Lu and Zhiqiang Yang
Project Location: Tongxiang, Zhejiang Province
Area: 23,000 sqm
Completion Time: 2020

濮院时尚工场坐落于濮院羊毛衫市场核心区，其前身濮院老物流园承载了濮院数十载商贸往来，也是濮院时尚产业发展壮大的见证者。为适应濮院时尚产业的创新发展，我们受邀对该园区进行改造升级，将其打造成为符合新一代创业人需要的时尚创业街区。升级后的时尚工场以花园式批零一体新型 MALL 的全新形象登场，成为濮院时尚新地标，也是目前濮院规格最高、时尚元素最强的专业市场之一。

Puyuan Fashion Park is located in the core area of Puyuan Sweater Market. Its predecessor, Puyuan Logistics Park, used to be a place for business and trade in decades. It is a witness to the development and growth of Puyuan's fashion industry. In order to adapt to the innovative development of Puyuan, the design team was commissioned to transform and upgrade the park, making it a fashionable entrepreneurial zone that meets the needs of a new generation of entrepreneurs. The upgraded park is given a new identity of a garden-style mall for wholesale and retail fashion business. It will become a new landmark of Puyuan and one of the most popular markets in the fashion industry.

经过深入分析和研究，我们提出以花园式批零一体新型 MALL 作为园区升级后的全新定位，打造集地标性建筑、网红打卡景点、直播摄影地、时尚展贸交易地、濮院时装周举办地、摩登青年消费首选地等标签于一体的创新型产业园区。

创新型产业园区的建筑设计承载着园区内文化人、创意人、技术人等复合群体的情感认知，需要营造文化认同，从而建立产业集聚的向心力。设计师保留原建筑主体结构，对建筑外立面再设计，打造个性化的公共空间，融入幻彩森林、春夏秋冬等空间主题，成功创造集体认同的场所精神，构建理想的生态型产业园区。

After in-depth analysis and research, the designer proposed to define the park after upgrading as a new garden-style mall for wholesale and retail fashion business, a landmark building, an Internet-famous place, a live broadcast site for Internet influencers, a fashion exhibition hall and trading place.

The architectural design of an innovative industrial park should meet the needs of different groups in the park: the professionals, the technicians, the businessmen, etc., and it is necessary to create a cultural identity. The designer decided to retain the main structure of the existing building, redesign the façade, create a personalized public space, and integrate different themes such as symphony forest, spring, summer, autumn and winter. A place with collective identity is successfully created.

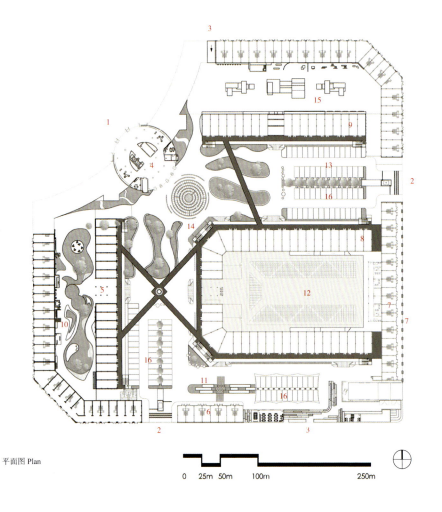

平面图 Plan

0   25m 50m   100m   250m

1 人行主入口 Pedestrian Main Entrance
2 车辆出入口 Vehicle Entrance
3 人行次入口 Pedestrian Secondary Entrance
4 1 号楼 No.1 Building
5 4 号楼 No.4 Building
6 5 号楼 No.5 Building
7 3-1 楼 No.3-1 Building
8 3 号楼 No.3 Building
9 2 号楼 No.2 Building
10 森之冬景区 Senzhidong Scenic Spot
11 森之春景区 Senzhichun Scenic Spot
12 森之心景区 Senzhixin Scenic Spot
13 森之夏景区 Senzhixia Scenic Spot
14 中心打卡区 Central Area
15 森之秋景区 Senzhiqiu Scenic Spot
16 停车场 Parking

整个园区以秩序的搭建来统领空间层次，以流线的梳理来弥合众多园区要素。新秩序的建立主要基于三点：

1. 基于园区内部：将功能空间和到达次序予以清晰的界定，并通过线性展示空间与点状租赁空间结合，使商业活动成为活跃空间的氛围核心。

2. 基于城市尺度：改造后的园区沿街，界面兼具古典的序列感和现代的形态质感。由半圆为主题衍生出的建筑外立面体系赋予园区灵活而优雅的印象，提升建筑在街区的辨识度，升级园区商业形象。

3. 基于环境生态：改造后的时尚工场以"幻彩森林"为主概念，以生态绿为核心元素，搭配春夏秋冬四季演变，勾勒出园区整体的生态氛围，将产业园区打造成森林花园，形成城市中的绿岛，实现生态化的升级改造。

The park is designed with a clear layout and streamlined circulation. A good order of spaces is established in three ways.

1. Based on the park itself: Functional areas and an arrival order are clearly defined. The linear display spaces are combined with scattered rental spaces to make commercial activities the core of the atmosphere of the park.

2. Based on an urban scale: Along the street outside the park, the scene has both a classical sense of sequence and a modern form and texture. The building façade system is inspired from semicircle, giving the park a flexible and elegant impression and an upgraded commercial image.

3. Based on the environment and ecology: The transformed park takes "symphony forest" as the main concept and ecology as the core element. The four seasons are used as design themes to define the overall ecological atmosphere of the park. In this way, an industrial park is turned into a forest/garden, forming a green island in the city and realizing urban ecological upgrading and transformation.

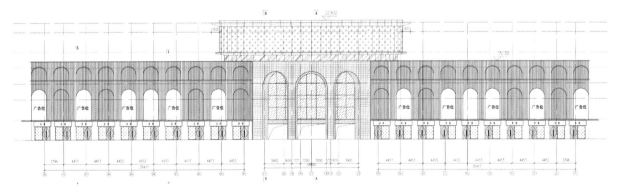

立面图 Elevation

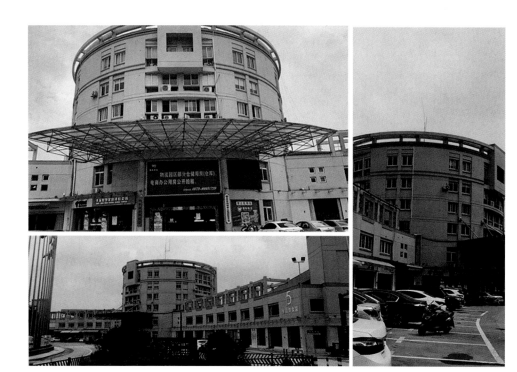

圆拱元素是世界上不同文化共同认同的一种典雅的象征。建筑外立面以格栅做成造型圆拱，作为装饰元素的单元之一。圆拱与格栅的排列既重复又充满变化，兼具古典的序列感和现代的形态质感。通过空间堆叠关系的变化，让其既是一个整体，又有着众多细节。视觉上也较为丰富，既是圆拱的重复排列，又是类似树干的切割形状的排列，与幻彩森林这个设计理念相契合。

局部前后交错的位置关系以及多层堆叠的形式，与古巴比伦的空中花园有着异曲同工之妙。为契合主题概念，设计以森林最基本的常态色——绿色，作为建筑外立面装饰的主色。

Arch is an elegant symbol that is recognized by different cultures around the world. The façade is clad by a grille system in the shape of arch, which serves as one of the decorative elements. The combination of the arch and the grille offers both a classical sense of sequence and a modern formal texture. The building, in the form of stacking structures, is an integrated whole with individual details. The architecture is visually rich, with the "arch and grill" system resembling cut trunks, keeping in line with the design concept of "symphony forest".

The partially staggered relationship and multi-layered stacking form are reminiscent of the hanging gardens of Babylon. To match the concept, the design uses the typical color of forest, green, as the main color for the exterior decoration of the building.

主楼六层作为园区主视觉窗口，其飞碟式顶层设有 360 度全景融合的空中展厅，周身被彩虹玻璃墙包围。以幻彩飞碟造型形成独特记忆点，将"五彩嘉兴"的寓意融入顶层的设计中，客商可以站在幻彩飞碟顶部，俯瞰濮院市场的霓裳风云。

主楼主体外立面升级后采用风铃幕墙包裹，如会呼吸的墙，增强了建筑与自然之间的互动，同时大幅提升了园区的时尚感形象。景观停车场别具一番风情，绚丽的色彩结合建筑连廊的造型，增强景点吸引力，打造网红打卡地标性景点，提升创意园的知名度。

The main building, consisting of six floors, serves as the main "window" of the park, with its UFO-shaped top floor featuring an air exhibition hall with 360-degree panoramic views surrounded by rainbow glass walls. The unique rainbow UFO shape creates a memorable landmark, with the meaning of "colorful Jiaxing", where visitors can stand on top of the "UFO" and overlook the bustling Puyuan market.

After the upgrade of the main building's façade, it is now wrapped in a wind chime curtain wall, which acts as a breathing wall that enhances the interaction between the building and nature, while significantly improving the park's fashionable image. The landscaped parking lot has a unique charm, with a combination of bright colors and the architectural form of a connecting corridor, which enhances the attraction of the place, creating a popular landmark for taking photos and improving visibility of the park.

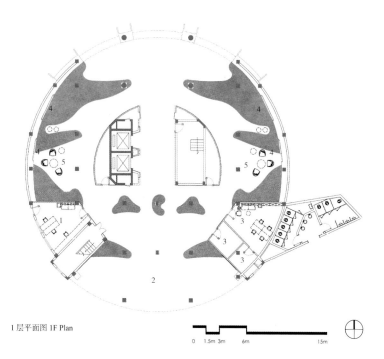

1 消防与监控室 Fire Control Room
2 大厅 Hall
3 服务用房 Service Room
4 广告位 Advertising
5 休息区 Rest Area

1 层平面图 1F Plan

0 1.5m 3m 6m   15m

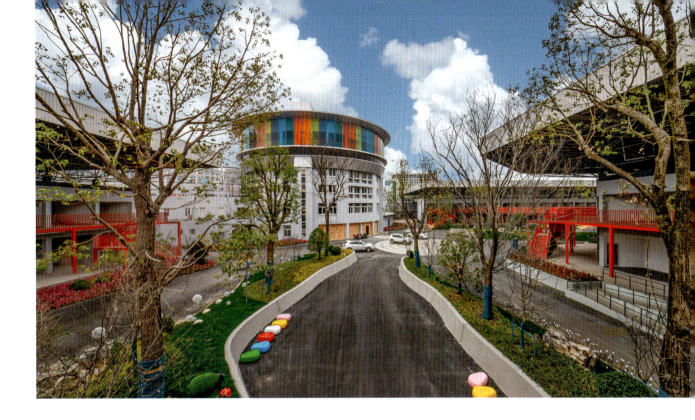

红色莫过于时尚工场内最亮眼的颜色。红色连廊连接着主楼之间的往来，也寓意濮院时尚产业的纽带在这里交织生长。

我们考虑到楼宇之间单独分离的空间结构，采用连廊的方式使其相互连接，通过一系列性能、承重、结构、施工等维度的反复讨论，确定钢结构多连体式连廊方案。通过连廊，外架楼梯将原有每户独立使用的二层改造成具备商业价值的一个整体商业空间。可将二层作为新商业进行出租运营，不仅增加了营业利润，还可大大促进整个综合体的商业活力，极大地提升综合体的可玩性、可逛性，亦可降低后期的一些招商运营难度。同时让流线更丰富，符合商业运营动线需求，与一层商业形成良好的互动。

Red is the brightest color in the park. The red corridor encourages exchanges between the main buildings, and also symbolizes that the ties of Puyuan's fashion industry are intertwined and grown here.

Considering the separate structures, the design team adopted the way of corridors to connect them to each other. Through repeated discussions on performance, load-bearing, structure and construction, the scheme of steel structure corridors was determined. Through the corridors, the external staircase transforms the original second floor into an overall commercial space. The second floor can be rented as a new business, which not only increases the operating profit, but also greatly promotes the commercial vitality of the entire complex. Meanwhile, it will also reduce difficulties of investment attractions in the later stage, facilitate free circulation required by commercial operation, and allow good interaction with the first floor business.

园区内主秀场面积约 3000 平方米，可容纳 1000 位观众。

在时尚工场内举办的每一场走秀活动，都将采用目前清晰度最高、透视效果最好的显示设备之一——冰屏。这是一种新型 LED 显示技术，看起来好像一组线状 LED 小灯组成的百叶窗，通透率最高可达 85%，极大提高了透视效果。观众站在理想距离观看时，画面就像悬浮于玻璃之上，给观众如梦如幻的沉浸式体验，带给观众非同寻常的时尚之旅！

濮院时尚工场改造在预算和施工时间双严控的前提下，没有采用大拆大建的建设模式，而是采取单元化，轻质化的加建方式，辅以景观手段处理外部空间。作为整改大于重建的既有建筑更新项目，改建的作用绝不仅限于功能的完善与形象的美化，而在于能提纯场所自身的特征，借由对空间格局的调动，达成感染人心的环境重塑，实现产业的聚集、转型和升级。

The main runway in the park covers an area of about 3,000 square meters, with a seating capacity for 1,000 spectators.

For every fashion show held here, one of the best devices available, the "ice screen", will be used. This is a new type of LED display technology that looks like a set of linear LED lights forming a shutter, with a maximum transparency of up to 85%, greatly enhancing the perspective effect. When viewed at an ideal distance, the picture appears to be suspended on the glass, offering the audience an immersive and dreamlike experience, bringing them on an extraordinary fashion journey!

The transformation of Puyuan Fashion Park is not based on a large-scale demolition and reconstruction model, but on a lightweight and modular approach, supplemented by landscape design to upgrade outdoor spaces, under strict control over budget and construction time. As a renewal project with the principle of improvement rather than reconstruction, the renovation not only improves functionality and enhances appearance, but also enhances the unique features of the place, achieving the transformation through adjusting the layout and realizing aggregation, transformation, and upgrading of the industry.

金沙书画院
JINSHA ART CENTER

2018 "永隆·星空杯" 江苏省室内设计大赛优秀奖
2020 金堂奖年度杰出样板房 / 售楼处空间设计奖
2020 她设计年度优秀空间设计奖

■

业主单位: 成都太行瑞宏房地产开发有限公司 - 金沙城·
　　　　云津观棠示范区
建筑设计: 吴耀隆
景观设计: 重庆创合园林设计有限公司
建设单位: 四川中辰城建装饰设计工程有限公司
项目地点: 成都市新津县
面积: 2000 平方米
完工时间: 2019

Client: Chengdu Taihang Ruihong Real Estate Development Co./
　　　　Yunjin Kwun Tong Model Area, Jinsha Town
Architectural Design: Yaolong Wu
Landscape Design: Chongqing Chuanghe Landscape Design Co.
Construction Company: Sichuan Zhongchen Chengjian
　　　　　　　　　　　Decoration Design & Engineering Co.
Project Location: Xinjin County, Chengdu, Sichuan Province
Area: 2,000 sqm
Completion Time: 2019

仁者乐山，智者乐水，山水赠诗人以灵气，诗人答山水以多情。文人墨客对新津这座 1400 余年古城有着独特的偏爱，本案的故事正式在这人杰地灵之地拉开序幕，在"金沙书画院"中徐徐上演。

A benevolent man delights in the mountains and a wise man delights in the water. The mountains and the water have given poets their aura and poets have answered them with affection. The story of this case officially opens in this land of great people and spirituality, and is being staged in the "Gold Sand Art Center".

成都新津金沙书画院由我们全程一体化设计，包括商业区的规划以及建筑、室内、软装、标识设计等。从设计到最后的施工落地建成，项目打破传统示范区的形式，以书画院的形式完美呈现了一个高标准，且结合科技创新设计、倡导低碳与环保的设计理念，实现可持续、多元化、复合功能的示范区。

书画院的主题来源于设计师对当地历史代表人物北宋宰相张商英及老四川文化和生活习惯进行深入挖掘。因此在新中式建筑、景观、室内空间中充分体现地域特色，将历史文人、诗意书画与自然、山、水、竹、石等元素结合，通过设计将美学融入空间中。在诗情画意之上，以水为灵性，以竹为精神，结合空间，诠释东方式儒雅隽永的意趣，匠心勾勒出人与自然和谐共处的理想空间。

我们将古人的智慧、精神、文化加以传承，发扬光大、勉励后人，让历史与当代对话、未来对话，构造自然美好的生活愿景。

The Chengdu Xinjin Gold Sand Painting and Calligraphy Academy was fully designed by the design team, including the planning and design of the commercial area, architecture, interior, soft decorations, and signage. From design to the final construction, the project breaks the traditional form of demonstration zones and presents a high-standard model of a painting and calligraphy academy. It combines innovative technological design, advocates for low-carbon and environmental protection design concepts, and achieves a sustainable, diversified, and composite functional demonstration zone.

The theme of the painting and calligraphy academy comes from the designer's in-depth exploration of local historical figures, such as Northern Song Dynasty Prime Minister Zhang Shangying, and the old Sichuan culture and living habits. Thus, the new Chinese-style architecture, landscape, and interior spaces fully reflect regional characteristics, integrating historical literati, poetic painting and calligraphy with natural elements such as mountains, water, bamboo, and stones, and integrating aesthetics into space through design. Above the artistic conception of poetry and painting, with water as the spirit and bamboo as the soul, combined with space, it interprets the elegant and lasting charm of the Eastern Confucianism aesthetics, and depicts the ideal space where human and nature coexist in harmony with ingenuity.

The designer inherits the wisdom, spirit, and culture of ancient people, encourages future generations, and lets history dialogue with the contemporary and the future to construct a beautiful vision of natural living.

项目作为一个多元化复合功能空间，将作为永久建筑的一部分被保留下来。设计上更多地运用低碳、环保节能材料，建筑与室内均采用了中式拼缝的手法工艺，大量采用新式科技材料竹钢与玻璃在建筑与室内的一体化设计上加以应用。建筑立面设计通过竹子衍变所带来的灵感，希望承载人文精神，赠予竹之节节于现在的我们。

项目景观以河流为主线贯通园林，打造四重景观体系：湖区景观、阳光绿轴、儿童主题区、社区小公园。用现代手法演绎五河交汇、人文贯通的浪漫古典主义，体现依水而居、步移景异、小中见大的东方意境空间。

As a multi-functional space, the project will be preserved as a part of the permanent building. The design makes more use of low-carbon, environmentally friendly and energy-saving materials, and the building and interior both adopt the technique of Chinese patchwork, and a lot of new technological materials (bamboo steel and glass)are used in the integrated design of the building and interior. The design of the building façade is inspired by the derivation of bamboo, hoping to carry the spirit of humanity and gift the bamboo to us now.

The landscape of the project takes the river as the main line through the garden, creating a four-fold landscape system: lake landscape, sunshine green axis, children's theme area, and community mini-park. The romantic classicalism of the intersection of five rivers and humanistic penetration is interpreted with modern techniques, reflecting the oriental mood space of living by water, moving scenery at every step and seeing the big in the small.

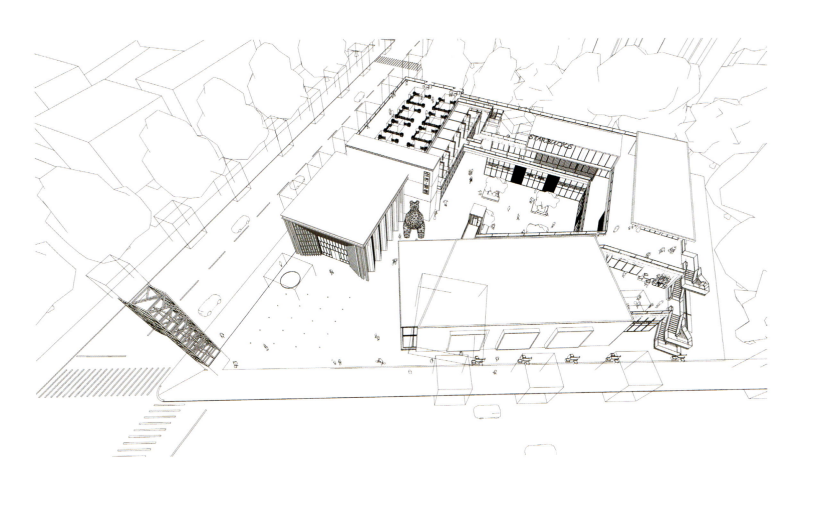

建筑地块 Building Plot

中庭镂空 The Atrium is Hollow

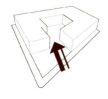

切主入口 Cut Main Entrance

切次入口 Tangent Entry

体块细分 Block Subdivision

通道连接 Channel Link

城市更新及建筑与室内一体化设计案例 URBAN RENEWAL, ARCHITECTURAL AND INTERIOR INTEGRATION DESIGN CASES　241

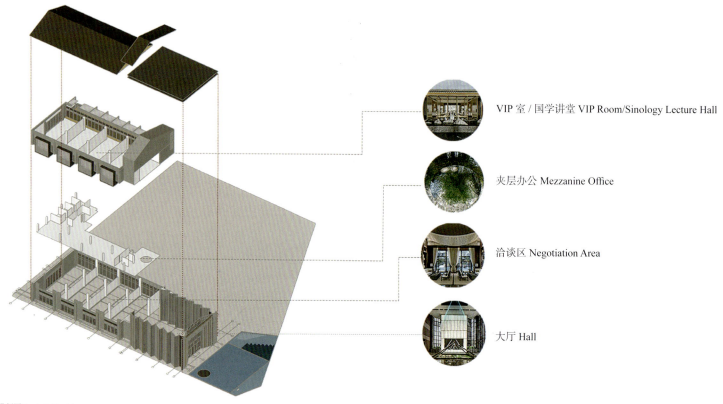

VIP 室 / 国学讲堂 VIP Room/Sinology Lecture Hall

夹层办公 Mezzanine Office

洽谈区 Negotiation Area

大厅 Hall

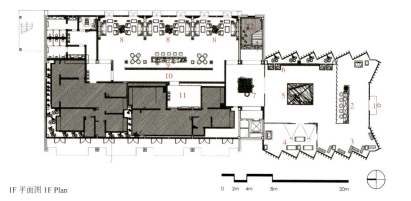

1F 平面图 1F Plan

0 2m 4m 8m 20m

1 主入口 Entrance
2 前厅接待 Front Hall Reception
3 品牌区 Brand Area
4 多媒体区 Multimedia Area
5 沙盘区 Sandbox Area
6 区位地图 Location Map
7 景观区 Landscape Area
8 洽谈区 Negotiation Area
9 水吧 Water Bar
10 时光长廊 Time Corridor
11 工法展示区 Construction Method Exhibition
12 鱼池水景 Fish Pond

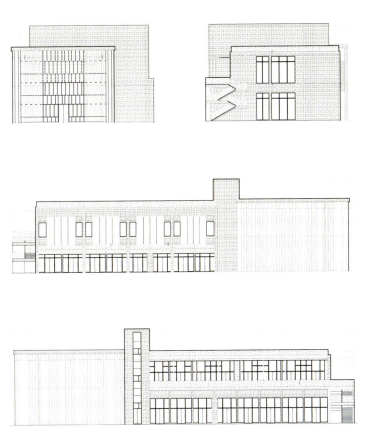

立面图 Elevations

进入室内一楼大厅，映入眼帘的是整面如水墨画般落地屏风，由上百张老四川的故事照片与玻璃材料相结合而成。设计延续中式拼缝组合，与建筑相呼应，透过光的折射，若隐若现。水纹肌理的大理石材，仿佛流动的水之韵律，充满浓厚的文化韵味，如同水墨画一般。这里是时空穿越的开始，进行着当代与历史的对话，娓娓道来新津的故事。

沙盘和多媒体区域有着高与低、大与小的空间转换。区别于传统销售中心，沙盘区域运用了诸多高科技与智能化技术，以三角形的艺术形态玻璃与钢结合，利用激光技术实现不同的灯光特效与喷雾的氛围烘托，让到访者有着非凡的视觉感受。竹钢背景下是清代著名画家张宗仓的《白云红叶图》，与老照片屏风形成不同时代的强烈对比，营造了一个穿梭古今的沉浸式体验场景。

Entering the interior on the ground floor, you are greeted by an entire ink painting-like floor-to-ceiling screen, made from hundreds of photographs of old Sichuan stories combined with glass materials. The design is a continuation of the architectural style, with a combination of Chinese stitching that echoes the building and appears through the refraction of light. The watery texture of the marble material resembles the rhythm of flowing water and is full of strong cultural flavor, as if it were an ink painting. This is where the time travel begins, where a dialogue between contemporary and history takes place and the story of Xinjin is told.

The sandbox and multimedia area has a spatial transformation between high and low, large and small. Unlike traditional sales centers, the sandbox area uses a number of high-tech and intelligent technologies, combining glass and steel in the form of triangular art forms and using laser technology to achieve different lighting effects and spray atmosphere to give visitors an extraordinary visual experience. Against the bamboo steel backdrop is the famous Qing Dynasty painter Zhang Zongcang's "White Clouds and Red Leaves", which forms a strong contrast with the old photo screens of different eras, creating an immersive experience that enables you to travel through the past and present.

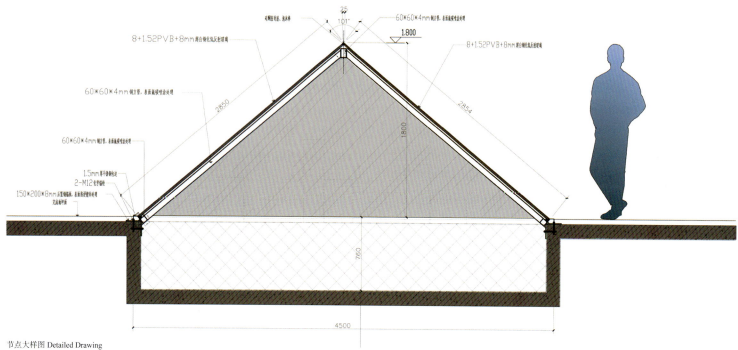

节点大样图 Detailed Drawing

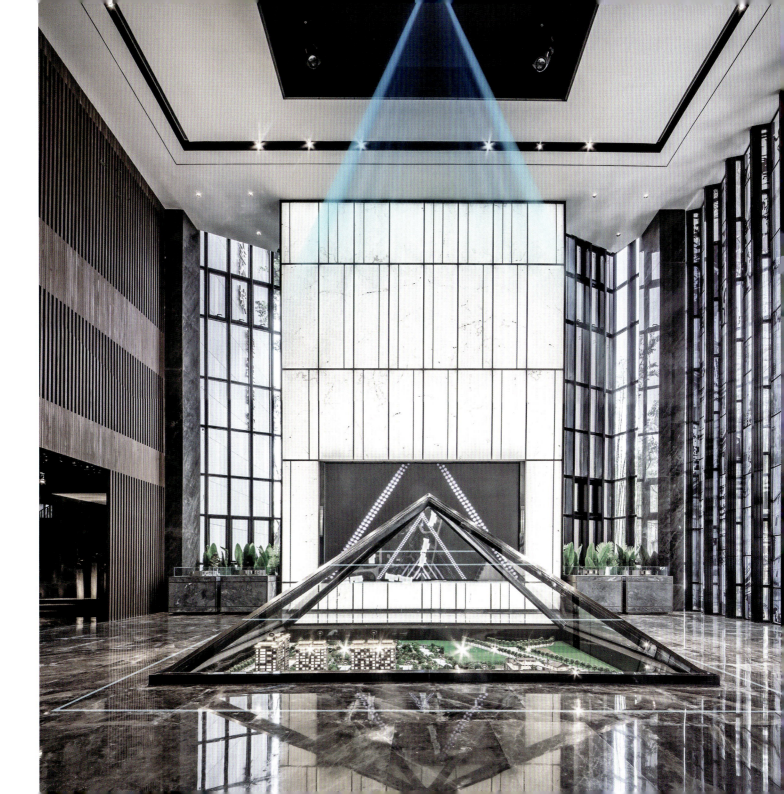

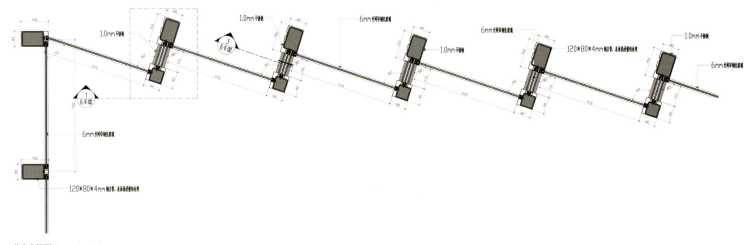

节点大样图 Detailed Drawing

室内景观曲折转换，收与放的关系，开与放的空间，天圆地方的设计，让人充满冥想。一组组生机勃勃的竹子，向圆穹生长，宛如要穿破天空；抬头有水纹灵动，由金属个锈钢板锤凿而成，仿佛悬浮于天与地之间，感受万物生灵。空间中灵动的光影斑斓，虚实之间，沉浸其中，不亦乐乎。

The interior landscape transforms in twists and turns, filling the room with meditation. A group of vibrant bamboos grows towards the round dome, as if to pierce the sky; looking up as if the water ripples spiritually, hammered by metal without steel plates, as if suspended between the sky and the earth, feeling the life of all things. The space is spirited with splashes of light and shadow, as between reality and illusion, and it is a pleasure to be immersed in it.

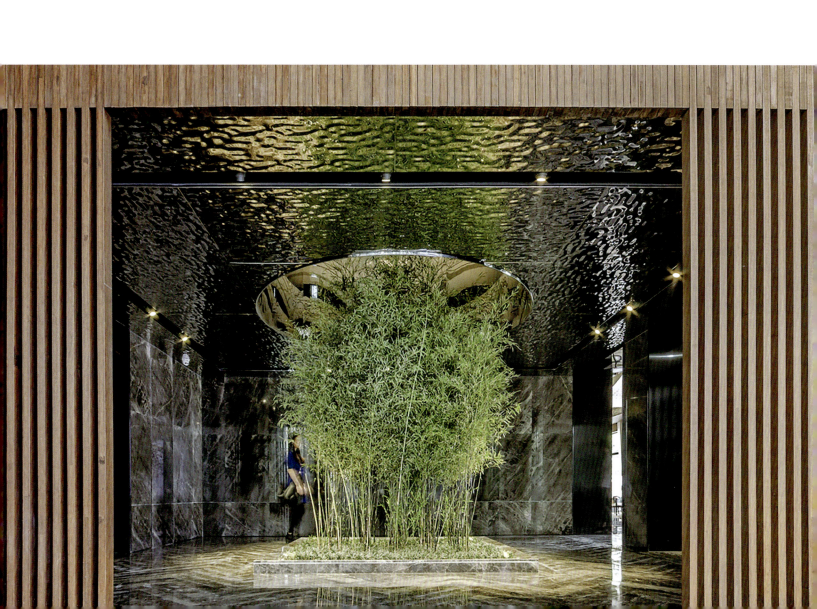

洽谈区的中央背景墙是北宋王希孟的《千里江山图》，画面细致入微，烟波浩渺的江河、层峦起伏的群山构成了一幅美妙的山水景象。行至空间中心为水吧区，它连接时光长廊与洽谈区，是一个可开可合的多功能组合空间。敞开式的空间处理方式打破了功能分区带来的边界感。顶部通过水滴模拟苍穹，与地面的静水山水呼应，充满仪式感，以隐喻的手法，营造出天空之城仙境梦幻般唯美场景。

The central backdrop of the negotiation area is a painting by Wang Ximeng of the Northern Song Dynasty, "A Thousand Miles of Rivers and Mountains", which is a detailed picture of a wonderful landscape scene with a vast river and rolling hills. At the center of the space is the water bar, which connects the time corridor with the negotiation area, a multi-functional combination space that can be opened or closed. The openness of the space breaks down the sense of boundaries caused by functional divisions. The top simulates the dome of the sky through water drops, echoing the still water of the landscape on the ground, full of ritual, and creating a beautiful scene like the dreamy wonderland of the city in the sky with a metaphorical debut.

效果图 Rendering

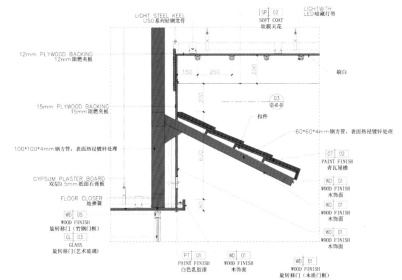

节点大样图 Detailed Drawing

步入二楼洽谈区，豁然开朗的大幅弧形卷轴长卷书法，带来震撼的视觉冲击，划破了原硬朗的空间。中卷衔接，浓重的中式韵味扑面而来，将传统文化元素直接展现于眼前。

行走于空间中，犹如画卷般被层层打开，人会不自觉地被周围的场景感染，激发内心深处的共鸣。

Stepping into the negotiation area on the second floor, a large curved scroll of calligraphy opens up, bringing a stunning visual impac. The middle scroll is articulated with a strong Chinese flavor, bringing the language of traditional culture directly to the forefront.

Walking through the space, like a scroll being opened up in layers, one is unconsciously infected by the surrounding scenes, inspiring a deep inner resonance.

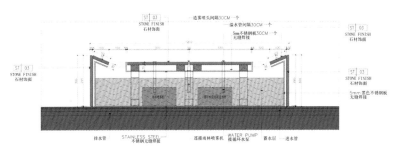

节点大样图 Detailed Drawing

二楼 VIP 洽谈室亦是可开可合、充满仪式感的空间，以闻风、听雨、见山、望水命名。儿童活动区域我们别出心裁地以国学讲堂的形式打造，让孩子们重温古人一起上课的画面。空间以中式对称的布局方式表达当代中国文人的精神和居住生活理念，木饰面为主要装饰材料，并以灯为介质辅以水墨画装饰，给空间中加以愉悦、轻松的沟通氛围。

从室内到建筑、景观，呈现出项目独特的东方式儒雅气质，我们独具匠心地勾勒出人文历史与自然和谐的理想空间。以书院的方式演绎清雅含蓄、端庄丰华的东方精神场所，以瞬间见永恒的诗性凝聚视觉，创造出充满生命仪式的"金沙书画院"。

The VIP negotiation room on the second floor is also a space that can be opened and closed, named as "the wind", "listening to the rain", "seeing the mountains" and "looking at the water". The children's activity area has been designed in the form of a Chinese lecture hall, allowing the children to relive the images of ancient people attending classes together. The space is laid out in a Chinese symmetrical way to express the spirit of contemporary Chinese literati. Wooden finishes are the main decorative material and are supplemented by ink paintings using lamps as a medium to give the space a pleasant and relaxed communication atmosphere.

From the interiors to the architecture and landscape, the project presents a unique oriental and elegant atmosphere, and the designers have uniquely outlined the ideal space where people and nature live in harmony. The project is an oriental spiritual place that is elegant and subtle, dignified and rich in the manner of a school, creating a "Sands Academy" full of life rituals with a poetic cohesive vision of eternity in an instant.

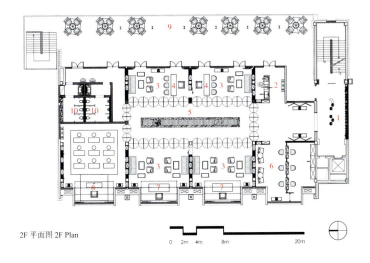

2F 平面图 2F Plan

0 2m 4m   8m   20m

1 景观区 Landscape Area
2 水吧 Water Bar
3 贵宾室 VIP Room
4 功夫茶 Kong Fu Tea
5 景观水池 Landscape Pool
6 收银区 Cashier
7 书画台 Painting Platform
8 儿童国学讲堂 Children's Sinology Lecture Hall
9 室外露台 Outdoor Terrace
10 洗手间 Toilet

中海·海上和集 POP-UP 体验店

CHINA OVERSEAS: THE GATHERING POP-UP STORE

XY+Z DESIGN 晰纹设计携手中海地产，将位于上海杨浦区海上海弘基休闲广场内一栋老建筑改造为全新的"中海·海上和集 POP-UP 体验店"空间，由吴耀隆先生与郭晰纹女士共同操刀完成，从设计方案的通过到整体改造落地完工仅短短一个月，实际施工时间仅 22 天，期间包含根据多方意见进行调整方案，项目仕时间和成本把控方面都面临了巨大的挑战。

XY+Z DESIGN Design cooperated with China Overseas Land & Investment Ltd. to transform an old building located in Hongji Leisure Plaza, Yangpu District, Shanghai into a new place called "The Gathering Pop-up Store", which was jointly completed by Mr. Yaolong Wu and Ms. Xiwen Guo. It took only one month from the design scheme to the completion of the overall transformation, and the actual construction time was only 22 days. During this period, adjustments were made based on multiple opinions, and the project faced significant challenges in terms of time and cost control.

■

业主单位：中海地产
建筑设计：吴耀隆
项目地点：中国上海
面积：60 平方米
完工时间：2022

Client: China Overseas Land & Investment Ltd.
Architectural Design: Yaolong wu
Project Location: Shanghai, China
Area: 60 sqm
Completion Time: 2022

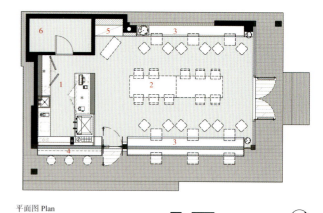

平面图 Plan

0 0.5m 1m   1.5m   5m

1 咖啡制作点单区 Order Area
2 多功能洽谈区 Multi-Functional Negotiation Area
3 卡座洽谈区 Booth Negotiation Area
4 吧台区 Bar Area
5 咖啡文创区 Coffee and Creative Area
6 储藏室 Storeroom

立面图 Elevation

"海上海"——由住宅、LOFT、商业三大物业形态及海上剧场、海上展馆、海上讲堂三大文化建筑组合而成的国际创意社区，主要建筑群体在近二十年的时间里依旧保持着活力，但部分独栋小型建筑由于暂停使用后逐步废弃，多年的变化累积，现有的空间结构混乱、外立面陈旧残破，如何在遵循原有基本外观结构的同时，使其恢复生机是本次设计的一大挑战。待改造建筑位于创意社区景观区域，一侧邻近沿街商业板块。考虑到这种独特的场地条件，设计既要考虑面向街道的立面与商业板块的协调性，又不能浪费面向景观区域的自然景观。

"The Gathering" is an international creative community composed of three major property types: residence, LOFT and commerce, including three cultural buildings: a theater, an exhibition hall and a lecture hall. The main building groups have maintained vitality in the past two decades, but some small single family buildings have been gradually abandoned after being suspended from use. Due to years of changes, the existing structure is chaotic and the façade is old and dilapidated. The design follows the original basic structure. XY+Z DESIGN needs to make it regain its vitality. The renovated building is located in the creative community landscape area, with one side adjacent to the commercial section along the street. Taking into account this unique site condition, the design not only considered the compatibility of the façade facing the street with the commercial blocks, but also did not waste the natural landscape facing the landscape area.

2022 年 10 月 8 日，"中海·海上和集 POP-UP 体验店"正式动工，开始为期 22 天的改造升级计划。最终将会呈现一个极具个性和创造特征，以恰当的分寸、极简的手法体现设计层次感与通透性的小型建筑作品。

空间仅 60 平方米，四周绿化植被丰富，视野开阔，在喧杂躁动的城市中心生长出一处"恬静淡然"的空间，在这里建筑、自然与人互动相融。没有约束，以轻松、简单、亲近自然的理念进行空间设计，回归人类自由的天性。

On October 8, 2022, the CHINA OVERSEAS: THE GATHERING POP-UP STORE project officially began, starting a 22-day renovation and upgrading plan. In the end, it presented "a small structure with great creativity", reflecting a good sense of permeability with appropriate minimalist approaches.

"The Gathering" is a store with only 60 square meters. Surrounded by well grown greenery, it has wonderful broad views. A tranquil place is created in the buzzy city center, where architecture and nature will interact with man. Here people have no constraints. The space is designed to make you relax in nature and recall an innate quality of human beings: yearning for freedom.

北侧外立面采用不锈钢镜面与镜面贴膜组合，空间的界限感一点一点消失在一片绿意之中。夜色下，温柔地散发着橘色的微光。时间，慢下了脚步。你，在看风景；风景，也在看你。

我们通过大面积折叠玻璃窗将室内延伸到室外，折窗下可两侧对坐。户外区树下光影斑驳，和煦的微风迎面吹拂，内外贯通，浑然一体。

The north façade adopts the combination of stainless steel and film. The mirrored surfaces reflect the surroundings, helping the building merge into the environment. Architecture and environment are integrated and interwoven.

The designer extends the interior to the exterior through a large area of folding glass windows, which provide seats on both sides. A play of light and shadow is created under the trees in the outdoor area. The gentle warm breeze that blows in connects the inside and outside naturally.

进入"中海·海上和集 POP-UP 体验店"内部，天花的设计手法表达了简洁与留白的意境。天花的斜顶设计在完美包揽了原有天花繁杂的钢筋结构的同时，形成舒适的美学体验。

水吧区采用现代亚克力材质，以分解、再组合的形式搭建出"中海·海上和集"标志，榫卯结构的图形创意被一次次重新解构。整体水吧台采用复合功能设计，日常兼顾小酒吧、咖啡软饮及简餐服务。

In "CHINA OVERSEAS: THE GATHERING POP-UP STORE", the ceiling is designed with the principle of simplicity and "blank-leaving" a technique highly valued in traditional Chinese paintings. The inclined ceiling perfectly covers the complicated steel structure of the existing ceiling, providing a comfortable visual appearance.

In the water bar area, modern acrylic materials are used to build the logo of "The Gathering". The tenon and mortise structure is deconstructed and reconstructed countless times. The bar is designed with multiple functions, providing wine, coffee and other drinks, as well as simple meals.

内侧利用大面积镜面不锈钢，本身就足够吸睛的标志灯光装置在镜面的作用下，进一步放大了交错、重复的秩序感，使整个空间得以延展。

黑色金属板饰面的护墙板与沿窗座位连成一体，空间功能合理交织且独立，自然光与暖色灯光安静地描述一副惬意画面。

Stainless steel panels are used on the inner side. The mirrored surface enhances the visual effect of the lit logo, which is already very eye-catching in itself, extending the entire space visually.

The dado with black metal plate facing is integrated with the seats along the window. Areas with different functions are independent yet interwoven at the same time. Natural light and warm artificial light are combined, creating a quiet, pleasant scene.

麦田梦境
THE DREAM IN THE RYE

2017 "永隆·星空间杯" 江苏省室内设计大赛银奖

2017 中国地产设计大奖中国优秀奖

2018 中国室内设计大奖赛入选奖

2019 意大利 A DESIGN 铜奖

2019 德国国家设计奖

■

业主单位: 中粮祥云置业南京有限公司 - 南京中粮祥云销
售中心

建筑设计: 吴耀隆

主要材料: 穿孔铝板、艺术漆、氧化镁板

项目地点: 江苏省南京市江宁区鹏山路 28 号

面积: 1850 平方米

完工时间: 2017

Client: COFCO Xiangyun Real Estate Nanjing Co /COFCO Xiangyun
Sales Center, Nanjing

Architectural Design: Yaolong Wu

Main Materials: Perforated aluminum panels, art paint, magnesium
oxide panels

Project Location: 28 Pengshan Road, Jiangning District, Nanjing,
Jiangsu Province

Area: 1,850 sqm

Completion Time: 2017

首先，这是一个"艺术殿堂"！其次，这是一个改造的销售中心项目。

由于是一个销售中心项目，所以注定了我们与"时间"站在了对立面。不管设计还是施工，大家一起想了很多优化办法，在确保合规性和成本控制最大化的前提下，以最快的速度来推进工作。

两个月的时间，从概念到深化再到落地……终于，"麦田梦境"如我们想象中的一般，完美地呈现在我们面前。

Firstly, it is a "temple of art"! Secondly, it is a renovated sales center project.

As it is a sales center project, we were destined to be on the opposite side of the fence from "time". Whether it was design or construction, we all worked together to come up with a number of optimizations to ensure maximum compliance and cost control, and to move forward as quickly as possible.

In two months' time, from conceptualization to deepening to implementation... finally, the Dream in the Rye was presented in front of us as we had imagined.

The original building was a little dull and lacking in character, but in order to respect the original building without destroying its appearance and to create an overall effect, we let the main façade wear a "mysterious veil"-white perforated aluminum panels-to weaken the original style. The uniform perforated aluminum panels raise the volume of the building to complete the overall appearance, creating a pure white image of the building during the day, while at night the lighting creates a different color effect to attract the attention of passers-by. In addition, the perforated aluminum panels are structured in such a way that the original curtain wall system is not disrupted, the structure is light and delicate, and the materials are environmentally friendly and renewable, which greatly reduces construction difficulties and project costs. The curtain wall also decorates the night sky with the Aurora Borealis, which also decorates your dreams and mine.

We kept as much of the curtain wall glass as possible, kept the aluminum panels to a certain size, and added a redesigned logo, so that the old and the new coexist but look like new life.

原建筑外观略显沉闷且个性不足，但本着尊重原建筑不破坏外观的原则，同时打造一个整体效果，我们让它的主立面穿上了一层"神秘面纱"——白色的穿孔铝板。统一的穿孔铝板弱化了原有的风格，并将建筑的体量进行了抬高，使整体外观形象更为完整，在日间形成一个纯白的建筑形象，而到晚间通过灯光营造出不同色彩的效果，吸引来往路人的注意力。另外，这个穿孔铝板的结构基本没有破坏原来的幕墙体系，结构轻盈细腻，材料环保可再生，大大降低了施工难度和项目成本。幕墙因与"极光"的相遇点缀了整片夜空，也装饰了你我的梦。

我们尽可能保留了更多的幕墙玻璃，把铝板面积控制在一定的范围内，加上再设计的 logo，新旧共存却又宛若新生。

我们创造了一个巨大的"蜂巢"，这是专属于孩子的"小沙龙"。每个六角空间都设置了软木垫，外部则用麻绳网进行防护，为孩子们提供了一个绝对安全的游玩环境；而可视的防护网设计也为家长们提供了实时的监护环境。我们希望勤劳的"小蜜蜂们"能在这里留下非常美好的回忆。

转折楼梯与水吧相连，独特的造型又为其增添了一份艺术感。通体金色的扶手，一步步指引着我们的路线。金色，是麦田的颜色，而我们将会踏着麦浪逐步走向远方。

还记得门厅的大片麦田吗？我们打通了一到三层的几跨楼板，在二层设置"麦田"的同时，也在三层设置了全透的可上人玻璃，形成了一个悬空的玻璃平台。在这里，你可以用上帝的视角俯视整片麦田，伴随着蒙德里安享受你的艺术人生。

设计不是一个单一体系。除了建筑以及室内改造，我们还做了标识标牌等一系列相关产品设计。每一个细节我们都精心挑选、精确把控，只为项目最终的完美亮相。

就像一场梦，梦里有麦田、有极光，有我心里所想的一切。

两个月，一天天看它逐渐长成我心中的模样。

这里，就是麦田梦境。

A large "beehive" has been created as a "mini salon" for the children. Each hexagonal space is cushioned with cork and protected by a twine net on the outside, providing a perfectly safe environment for children to play in, while the visible netting provides a real-time supervision environment for parents. We hope that the hard-working "bees" will make great memories here.

The unique shape of the winding staircase that connects to the water bar adds another artistic touch. The golden handrail guides our way step by step. Gold is the color of the rye, and we will be stepping on the waves of wheat to the far side.

Remember the large rye in the foyer? We have opened up several floors from the first to the third, creating a suspended glass terrace on the second floor with a "rye" and a fully permeable glass roof on the third floor. From here you can look down on the entire field of wheat from the perspective of God and enjoy your artistic life with Mondrian.

Design is not a single system. In addition to the architectural and interior renovations, we have also designed a range of related products such as signage. Every detail was carefully selected and precisely controlled for the final presentation of the project.

It was like a dream, a dream with wheat fields, aurora borealis and everything I had in mind.

For two months, day by day, I watched it grow into the shape I had in mind.

This is the dream in the rye.

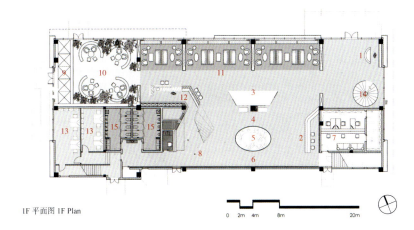

1F 平面图 1F Plan

1 门厅接待 Foyer/reception
2 销控台 Sales Control Station
3 精品展示区 Boutique Exhibition
4 户型沙盘 House Model
5 社区沙盘 Neighborhood Model
6 区位地图 Location Map
7 财务办公室 Finance Office
8 宣传展示长廊 Publicity Gallery
9 玩具屋 Dollhouse
10 沙龙区 Salon Area
11 洽谈区 Negotiation Area
12 水吧 Water Bar
13 办公室 Office
14 楼梯 Staircase
15 洗手间 Toilet

0 2m 4m 8m 20m

2F 平面图 2F Plan

1 品牌性能展示区 Brand Performance Display
2 体验互动区 Experience Interaction Area
3 商务主题贵宾室 Business Theme VIP
4 家庭主题贵宾室 Family Theme VIP
5 服务吧台 Service & Bar
6 室外冥想平台 Outdoor Meditation Platform
7 艺术展示区 Art Exhibition Area

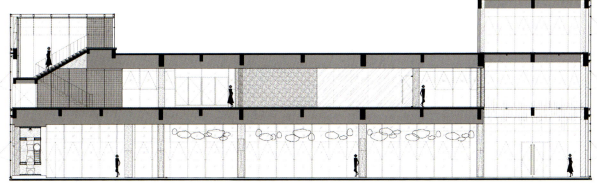

立面图 Elevation

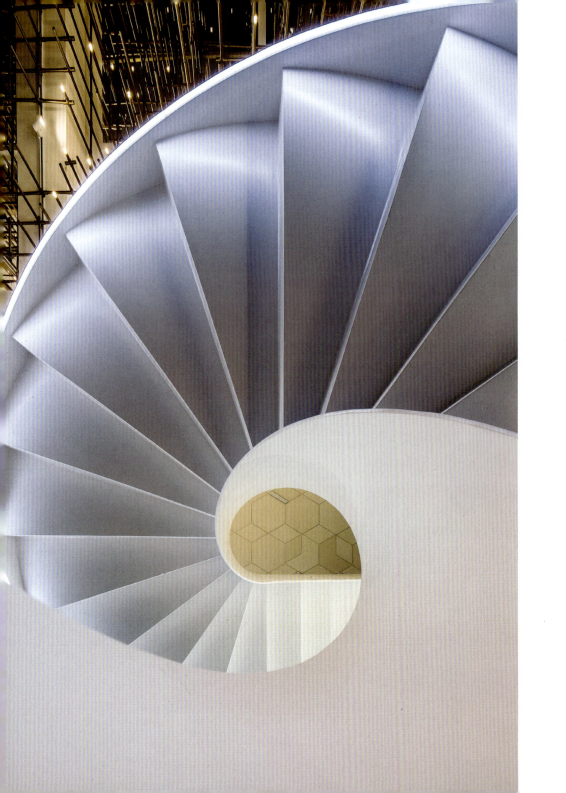

旋转楼梯是整个空间的"精神堡垒"，纯白无立柱的设计更加彰显了中粮的艺术气质。拾级而上，你能看到滚滚麦浪以及它的守护者——乐高沉思者。

The revolving staircase is the "spiritual fortress" of the entire space, and the pure white design without columns further highlights COFCO's artistic temperament. As you walk up the stairs, you can see the rolling waves of wheat and its guardian, the LEGO Contemplator.

随着地面光纤的引导，你会来到销控区。空间的"艺术气质"淡化了整体的商业氛围，这里，仿佛更像一个艺术馆。我们把"中粮"重构、再生，置于墙面、吊于天花或制成屏风，麦田的元素贯穿整个空间。麦，即是粮。

这是门厅，也是我们的"艺术殿堂"。墙面通体纯白背光，抬头，你能看到大片麦田犹如繁星点点照亮了整个空间；低头，又好像置于宇宙之中，艺术漆与光纤的巧妙融合，带你进入不一样的时空。

As the floor is guided by fiber optics, you will come to the sales and control area. The "artistic quality" of the space dilutes the overall commercial atmosphere, making it feel more like an art gallery. We have reconstructed and regenerated "COFCO", placing it on the wall, hanging it from the ceiling or making it into a screen, with elements of wheat fields running throughout the space. The wheat is the grain.

This is the foyer and our "art hall". The walls are backlit in pure white. Looking up, you can see a large "wheat field" like stars lighting up the whole space; looking down, it is like being placed in the universe, with the clever fusion of art paint and fiber optics, taking you into a different time and space.

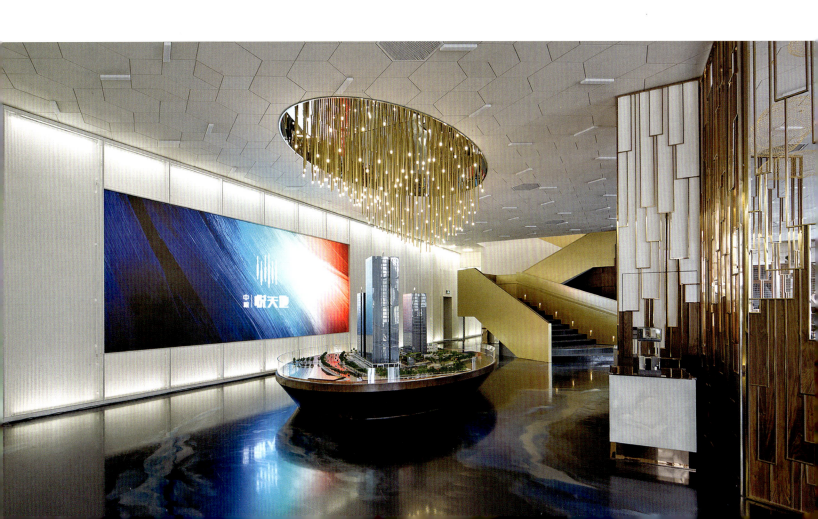

一个屏风，两个区域。洽谈区选用了北欧风格的家具搭配木质书架，打造一个舒适自然的交流环境。天花使用氧化镁板铺满整个区域，契合中粮 logo 的六边形元素，简洁又不失美感。

One screen, two areas. In the negotiation area,Scandinavian-style furniture and wooden bookshelves are used to create a comfortable and natural environment for communication. The ceiling uses magnesium oxide panels to cover the entire area, fitting the hexagonal elements of the COFCO logo, simple yet aesthetically pleasing.

二层的独立 VIP 是对麦田的延续，摆脱传统固有的设计模式，给它注入新的生命力。将凡·高丰收系列油画高清喷印在无纺布并贴于墙上，由顶上的防眩目轨道灯照亮整个空间，从地毯到墙壁，仿佛真的身处麦田之中，感受到阵阵麦香。

家庭 VIP 则继续引入"极光"的概念，天花上大胆使用绢布和灯光来模拟"极光"的效果。飘逸灵动，看一眼即会爱上，就连北极熊一家也在这里驻足。立面上运用流线型弧线将书架和卡座置于其中，希望赋予其情感和个性，整个空间就像童话一般充满趣味，相信来这里的每一个人都能有不一样的感受。

The separate VIP room on the second floor is a continuation of the wheat field, breaking away from the traditional design pattern and breathing new life into it. Van Gogh's Harvest series oil paintings are spray-printed in high definition on the walls, and the space is illuminated by dazzle-proof track lighting from the ceiling, from the carpet to the walls, as if you were actually in a wheat field, feeling the fragrance of wheat.

For the Family VIP room, the concept of "aurora" continues with the use of silk fabric and lighting on the ceiling to simulate the "aurora" effect. Even the polar bear family has stopped by. The façade uses streamlined curves to place bookshelves and card holders in it, hoping to give it emotion and personality, and the whole space is like a fairy tale full of fun.

在通往三楼的转角楼梯上，我们定义了一场"艺术之旅"。一条"框架长廊"截取了世界名画的部分画面贴于架身内侧并给予了新的定义。由此，一阶一阶形成了一条充满文艺气息的"艺术长廊"。置身其中，慢慢感受画中的每一个故事。而在它的对面，有另一座楼梯，纯白，舍弃了一切装饰和色彩。因为，有"极光"就足够了。白天，它是不入一丝尘埃的至净之地；夜晚，它身披彩霞，变成虚空的魔幻世界。

On the corner staircase leading to the third floor, we have defined an "artistic journey". A "framed gallery" takes some of the world's most famous paintings and places them on the inside of the frame, giving them a new definition. The result is a literary "art gallery", step by step. You can feel the story of each painting as you walk through it. Opposite it, there is another staircase, pure white, without any decoration or color, because the Aurora Borealis is enough. By day, it is a place of purity without a trace of dust; by night, it is a magical world of color and emptiness.

三

设计之外

III. BEYOND DESIGN

天然玉石遇见光 / 自然光系列
ORIGINAL FURNITURE

北极星的畅想
ORIGINAL PRODUCTS

致敬建筑大师
SALUTE TO ARCHITECTURE

中国传统再造
CHINESE TRADITION REINVENTION

关于 XY+Z 标志
ABOUT XY+Z LOGO

2021 年, 我与先生吴耀隆、平面设计师好友高斌女士、美学生活家好友兼甲方康康女士, 四位来自不同领域的设计师联合创办了"宇宙奇迹"中国原创跨界品牌, 提供原创跨界家具、珠宝产品, 以及专业室内软装设计及采购服务。三维的建筑、室内空间, 抑或是二维的平面创意, 不同视角延伸出设计的不同维度, 构建起截然不同的产品逻辑。我们认为, 多元文化带来感性思考, 专业场域催生理性度量, 加上无限创意的碰撞与结合, 才能诞生出称之为奇迹的创新作品。未来, 我们将邀请更多不同专业的设计师合作, 逐步推出更多产品, 与更多人一起发现中国原创设计之美。

产品、珠宝设计虽然是独立的专业, 但我同样觉得不同的角度和立场能够带来独特的创意和惊喜。他们其实是密不可分的。比如高迪, 比如扎哈。他们做规划、建筑、室内、产品, 甚至珠宝。我很喜欢他们, 随之而来, 他们也对我有较大影响。一个项目, 由一个总设计师负责理念, 会使整个项目由始至终地统一, 因为这样才能完整表达自己的作品。对我来说, 每个作品都应该是完整的, 有灵魂的, 有生命的。

这大概就是"宇宙奇迹"诞生的主要原因, 在这里你可以遇见那个更完整的我, 更多面的我, 更真实的我。

关于我对珠宝的喜爱这个事情, 看到这里相信你应该也已经感受到了。聚会时遭遇撞衫, 或是发现刚认识的朋友戴了和你同款的婚戒, 都不知道谁和谁是夫妻了……这种状况比比皆是。回归到初衷, 我做珠宝服饰设计也许就是为了骨子里那股"与众不同"。但你越是了解, 就越能发现跨界设计的难度及魅力。

设计师的审美和需求, 其实是比较"奇怪"的。我先生常说, 我发展珠宝服饰设计, 其实最主要是为我自己服务。这点我必须承认。笑。

跨界作品

CROSS-BORDER WORKS

In 2021, four designers from different fields—Ms. Gao Bin (a graphic designer), Ms. Kang Kang (an expert in life aesthetics, and also the client), Mr. Yaolong Wu (my husband), and I—co-founded the cross-border brand of XY+Z DESIGN, providing original cross-border furniture, jewelry products, as well as professional soft decoration in interior design and procurement services. Three-dimensional architecture and interior space and two-dimensional graphics provide different perspectives, extending dimensions of design with completely new design logic. We believe that multiculturalism brings perceptual thinking and expertise in different fields ensures rational decisions. The collision of infinite creative ideas will give birth to innovative works that can be called miracles. In the future, we will cooperate with more designers from different fields, gradually launch more products, and discover the beauty of Chinese original design with more people.

Although product and jewelry design are independent majors, I also feel that different perspectives and positions can bring unique creativity and surprises. They are actually inseparable. For example, Gaudí and Zaha. They are involved in planning, architecture, interior design, product design, and even jewelry design. I like them very much, and they also have a great influence on me. A project, with a chief designer in charge of the concept, will unify the whole project from beginning to end, because only in this way can one's design be fully expressed. For me, every project should be complete; it has its soul and life.

This is probably the main reason for the birth of the XY+Z DESIGN. Here you can meet a complete and real me.

My love for jewelry. I believe you should have felt this when you read this article. Perhaps you once met someone wearing the same dress as yours at the party, or found that a friend you just met was wearing the same wedding ring as yours, as if you are a couple! This is a commonly seen situation. As for the original intention of my design career, I think I started designing jewelry and clothes just to be "different" —an idea deeply rooted in my mind. But the more you understand, the more you will find the difficulty and charm of cross-border design.

The aesthetics and needs of designers are actually quite "strange". My husband often says that when I develop jewelry and clothing design, in fact, the most important thing is to serve myself. I must admit this. (Laugh.)

天然玉石遇见光 / 自然光系列
ORIGINAL FURNITURE

发光玉石桌
Luminous Jade Table

宇宙奇迹板材的诞生

机缘巧合下，一种新型自然材料——火山岩，进入了吴耀隆老师的视线。火山岩，这是由岩浆喷射地表或侵入地壳冷却凝固形成的神秘岩石，其形态天然，具有大小不一和散布不均的气孔。然而也正是这种特殊的形态与结构，引发了材料硬度问题，硬度不足让它在建筑与室内项目中实际运用困难重重。

在寻求解决方案的过程中，吴耀隆老师遇见了拥有超薄石材切割工艺的合作伙伴——颜界。这启发了吴耀隆老师大胆提出火山岩是否可以与其他材料复合。经过多轮研究尝试，火山岩超薄石片与玻璃覆合，成功提高了硬度。从此，一种如同置身于浩瀚宇宙的新型材料应运而生。

只有 0.3 厘米厚的火山岩薄片与玻璃覆合后的样品透过自然光投射出斑驳星光，在吴耀隆老师身边的女儿吴若可对其充满着好奇，不停追问父亲关于它的一切，而孩子无心的一句"这简直是宇宙的奇迹"一下击中了他的内心，就此，这款材料有了新的名称"宇宙奇迹"。

A New Material: XY+Z Board

By chance, a new type of natural material—volcanic rock—came into the sight of Professor Yaolong Wu. Volcanic rock is a mysterious rock formed by magma ejecting on the surface of the earth or invading into the earth's crust, cooling and solidifying. Its natural form, size, and uneven distribution of pores vary. However, it is precisely this special form and structure that leads to hardness issues. Insufficient hardness makes it difficult to apply in architectural and interior projects.

In the process of seeking a solution, Professor Wu met his partner Jie Yan, who has good experience in ultra-thin stone cutting techniques. This inspired Professor Wu to boldly propose whether volcanic rock can be used to make composite materials. After multiple experiments and attempts, he succeeded in combining ultra-thin volcanic rock fragments with glass, greatly improving the hardness. A new type of material emerged, which feels like being in the vast universe.

The composite material, which is only 0.3cm thin, emits speckled starlight under natural light. Ruoke Wu, the daughter of Professor Wu, was curious about it and kept asking her father about everything about it. The girl's unintentional words, "this is simply a miracle of the universe", hit his heart. As a result, this material got its name, "XY+Z".

长岛——多功能发光玉石

Multi-functional Luminous Jade Series: Long Island

海水泛起白色的泡沫，

岛上的人在等着烟火，

时间循环困住了四季，

我走进了别人的梦境。

天然原石，不拼不染，每座长岛纹理如高山流水，皆为独一。我们为"长岛"
设计了四种功能：茶盘／酒盘、食器、氛围灯和艺术摆件。

茶盘／酒盘：天然白云玉石纹理的独特美感，呈现出月光般细腻、温柔的画面。
透过玉石的微光，清澈、柔和，更加突出了茶汤或酒色的净、和、淡，更可
搭配创意杯垫"四季"组合使用。

食器：天然原石，可擦拭清洁，可直接接触食物。日常可以做餐厅摆盘，招
待、自享，独特纹理与柔光呈现食物最诱人的状态。

氛围灯：透过天然玉石的光散发出独特且柔和的美感。随意放置，多种光色
可调，亮度随心切换。

艺术摆件：搭配"四季"，自由组合，发现您的无限创意。

产品组合配件

四季——多功能亚克力摆件、玉石杯垫

以春、夏、秋、冬之四季为灵感，我们用心寻找与之灵魂相契的天然原石，
春之丛林、夏之深海、秋之大地、冬之冰川，我们为您收集人间四季，岁月
坦然，日落很慢，因果很淡。

极简的设计、高品质的制造工艺及深厚的文化内涵，让长岛成为品茶、用餐、
工作、阅读、休闲等多种场景的温暖陪伴，也期待您与我们共同解锁更多新
鲜有趣的创意玩法。

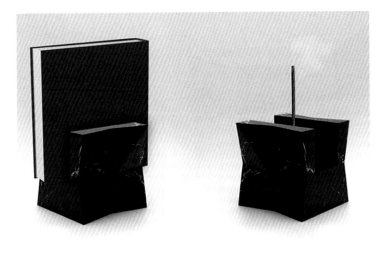

The sea water is suffused with white foam.

The people on the island are waiting for fireworks.

The time cycle has trapped the four seasons.

I walked into someone else's dream.

Natural raw jade, neither pieced nor dyed. Each "Long Island" has its unique texture. "Long Island" is a device designed with four functions: tea or wine tray, food utensil, lamp for ambient light, and artistic ornament.

Tea or wine tray: The unique texture of the natural jade presents a delicate and gentle image like moonlight. With the faint light of the jade, being clear and gentle, the pure texture of the tea or wine will be emphasized. It can be paired with the creative coasters designed with the theme of four seasons.

Food utensil: Made of natural raw material, it can be wiped and cleaned, and can come into direct contact with food. Daily dining can be displayed/served/enjoyed, with unique textures and soft light presenting the most enticing state of food.

Lamp for ambient light: The light emitted through the natural jade shows a unique charm. Light colors and brightness can be switched freely.

Artistic ornament: With the "four seasons" coasters you are allowed to discover your infinite creativity.

PRODUCT ACCESSORIES

"Four seasons" coasters: Multi-functional acrylic ornaments/jade coasters

Taking inspiration from the four seasons, we searched for natural raw jade that align with the souls of spring, summer, autumn and winter. The jungle of spring, the deep sea of summer, the land of autumn, and the glaciers of winter... we tried to present the essence of the four seasons.

Minimalist design, high-quality manufacturing techniques, and profound cultural connotations make "Long Island" a good companion for various occasions such as tea tasting, working, reading, and recreation. We look forward to you discovering more interesting functions of it.

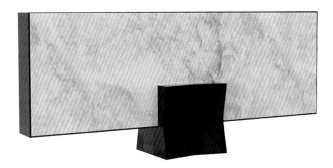

月光——多功能发光玉石

Multi-functional Luminous Jade Series: Moonlight

就像这世上找不到两轮相同的月亮，精选天然大理石与玻璃复合，让每一枚月光都是不可复制的孤品。

我们为"月光"设计了四种功能：茶盘或酒盘、花器、氛围灯和艺术摆件。

茶盘或酒盘：天然白云玉石纹理的独特美感，呈现出月光般细腻、温柔的画面。透过玉石的微光，清澈、柔和，更加突出了茶汤或酒色的净、和、淡。

花器：取出月光，让喜爱的花在造型底座上小"憩"。一盏茶，一束花，一个清幽的午后，便是岁月静好。

氛围灯：透过天然玉石的光散发出独特且柔和的美感。多种光色可调，随心切换。

艺术摆件：搭配创意底座，自由组合，发现您的无限创意。

产品组合配件

1. 山间——树脂摆件

通过造型刻画"蓬莱""方丈""瀛洲"三座神山，以模仿仙境，渐变墨色晕染如画的禅意世界。

2. 憩——多功能亚克力盘

以天圆地方为设计灵感，"憩"是"月光"憩栖之所。单独使用，可作为创意花器、置物盘、点心盒等，更多玩法等待充满创意的您来发掘。

极简的设计、高品质的制造工艺及深厚的文化内涵，让月光成为品茶、工作、阅读、休闲等多种场景的温暖陪伴，也期待您与我们共同解锁更多新鲜有趣的创意玩法。

Just like there are no two identical moons in this world, we carefully selected a natural marble and glass composite material to make each "Moonlight" an irreplaceable product.

"Moonlight" is a device designed with four functions: tea or wine tray, flower vessel, lamp for ambient light, and artistic ornament.

Tea or wine tray: The unique texture of the natural jade presents a delicate and gentle image like moonlight. With the faint light of the jade, being clear and gentle, the pure texture of the tea or wine will be emphasized.

Flower vessel: Take out the "Moonlight" and let your favorite flowers rest on its base. A cup of tea, a bouquet of flowers, a serene afternoon, a tranquil time...

Lamp for ambient light: The light emitted through the natural jade shows a unique charm. Multiple light colors are provided for selection and can be switched freely.

Artistic ornament: The creative base allows for opportunities to discover your infinite creativity.

PRODUCT ACCESSORIES

1. Mountains: Resin ornaments

By imitating the three sacred mountains of "Penglai", "Fangzhang", and "Yingzhou", we aim to create a fairyland and a picturesque Zen world.

2. The resting base: Multi-functional acrylic plate

Taking the "round Heaven and square Earth" as design inspiration, the "resting base" is the resting place for the "Moonlight". When used alone, it can serve as creative flower ware, storage tray, pastry box, etc., and more functions are waiting for you to explore.

Minimalist design, high-quality manufacturing techniques, and profound cultural connotations make "Moonlight" a good companion for various occasions such as tea tasting, working, reading, and recreation. We look forward to you discovering more interesting functions of it.

北极星的畅想
ORIGINAL PRODUCTS

■

北极星台盆

Polaris Basin

将北极星屋顶原型的灵感结合多彩亚克力及质感金属不锈钢，绚丽的北极星台盆应运而生，成为空间里的一抹亮色。

Combining the inspiration of the prototype of the Polaris roof with colorful acrylic and textured metal stainless steel, the magnificent Polari door basin emerged as a bright color in the space.

北极星灯具

Polaris Lamp

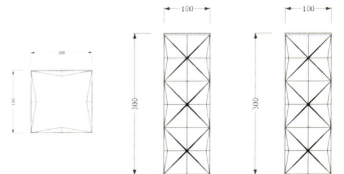

灯具表面独特的屋顶造型与透光性良好的亚克力结合，四射的星光璀璨闪耀，与珠宝的特质不谋而合。宇宙奇迹盒，作为珠宝北极星系列延伸产品，源自建筑屋顶结构的灵感与多彩亚克力的跨界结合，宇宙奇迹盒的诞生也是宇宙奇迹系列独特跨界融合理念的一个经典代表。

The unique roof shape on the surface of the lamp, combined with acrylic with good transparency, and the sparkling star light, coincides with the characteristics of jewelry. The XY+Z Box is an extension product of the jewelry Polari series. Inspired by the architectural roof structure and the cross-border combination of colorful acrylic, the birth of the XY+Z Box is also a classic representative of the unique cross-border integration concept of the XY+Z DESIGN series.

北极星五金

Polaris Door Handle

我们将北极星系列建筑屋顶原型的灵感与不同领域的产品进行跨界的融合延伸，与众不同的外观设计，结合金属不锈钢材质，于是诞生了北极星门把手。

We have integrated and extended the inspiration of the Polari series building roof prototype with products from different fields, creating a unique exterior design combined with metal stainless steel materials, resulting in the birth of the Polari door handle.

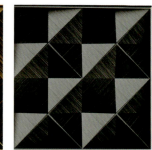

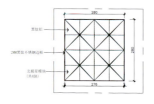
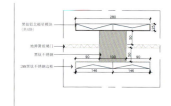

产品平面尺寸图 Product Specifications

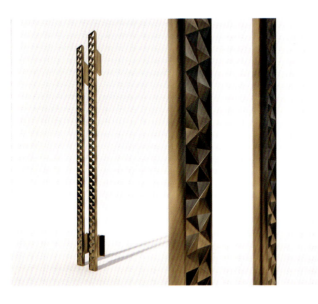

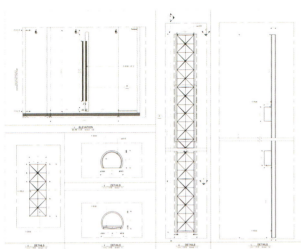

产品平面尺寸图 Product Specifications

北极星珠宝

Polaris Jewelry

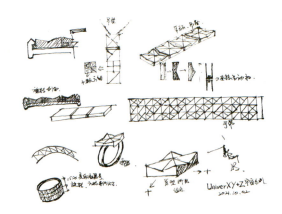

以建筑屋顶为灵感来源，同时解读中国折纸游戏，形成"北极星"。

并以单粒北极星为主要元素，进行不同的组合，构成不同的珠宝产品。

Using the roof of the building as inspiration, while interpreting the Chinese origami game to form the "Polaris".

And the single Polaris as the main element, different combinations, constitute different jewelry products.

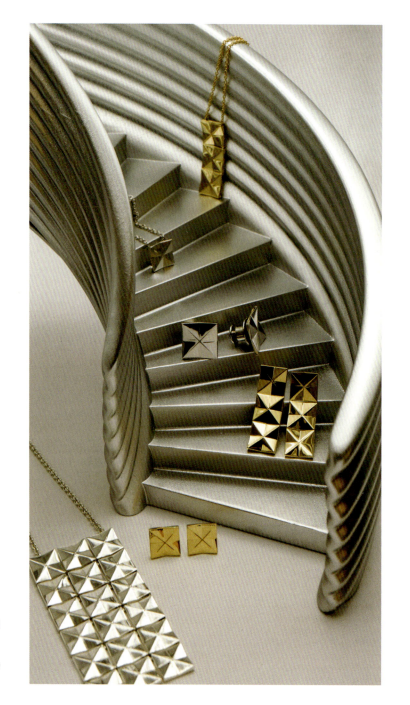

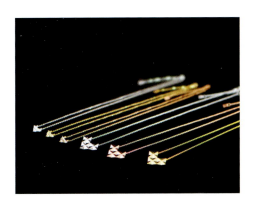

致敬建筑大师
SALUTE TO ARCHITECTURE

■

致敬高迪项链

Necklace: Paying Tribute To Gaudi

跨界系列——其实这个系列才是宇宙奇迹珠宝的核心。因为作为一个珠宝品牌而言，我们并不是一个专业玩家。建筑对人而言，是"庞然大物"，是只可进入而不可移动的。但是，跨界建筑珠宝的出现，打破了这个限制。多元化的交互，不同领域的设计师、艺术家的不同视角，带来意想不到的结论。"宇"众不同的设计作品及生活体验也由此产生。

吴耀隆老师："对我来说，这些大师的作品不仅代表了建筑的奇迹，也象征着无限的创新和勇气。他们的作品让人感到震撼，也激发了我对设计的热爱和追求。每一次看到这些作品，都会被其精妙的设计和无尽的可能性所打动。我会被他们对于形状、色彩、光影以及空间运用的技巧所吸引，并从他们的作品中汲取灵感和启发。不受传统束缚的勇气和极度投入的专注是建筑师一生追随的脚步。"

Cross-border series. In fact, this series is the core of XY+Z jewelry. We are not professional jewelry designers. For humans, architecture is a "giant object" that can only be accessed and cannot be moved. However, the emergence of cross-border "architectural jewelry" breaks this limitation. Diverse interactions, designers from different fields, and artists with different perspectives bring unexpected conclusions. The diverse products and life experiences have also emerged from this.

Yaolong Wu said: "For me, the works of these masters not only represent the miracles of architecture, but symbolize infinite innovation and courage. Their works are shocking and inspire my love and pursuit of design. Every time I see these works, I am moved by their exquisite design and endless possibilities. I will be attracted by their skills in shape, color, light and shadow, and space, and will draw inspiration from their works. The courage to be free from traditional constraints and the extreme dedication are the footsteps that architects should follow throughout their lives."

致敬路易斯·康：韦恩堡项链

Necklace: Paying Tribute to Fort Wayne by Louis Kahn

致敬路易斯·康：孟加拉国议会中心项链

Necklace: Paying Tribute to Bangladesh Parliamentary Center by Louis Kahn

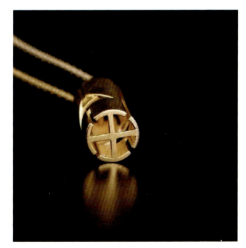

中国传统再造
CHINESE TRADITION REINVENTION

我的先生吴耀隆，同时也是宇宙奇迹品牌主理人，是一位有着强烈民族使命感的建筑师。凭借深厚的文化底蕴与敏锐的设计洞察力，在他的主导下，我们开始将中式传统建筑之精髓融入国风珠宝设计，打造出独具魅力的新国风系列。他深入挖掘中国传统文化的魅力，将古老的建筑纹样与现代家具及珠宝工艺完美结合，传承东方美学的精髓，赋予中式浪漫新的时代内涵。

中式传统建筑中的窗花、飞檐、斗拱等元素焕发新生，化作熠熠生辉的家具及珠宝饰品。亚克力、3D 打印、贵金属与珍稀玉石，将传统纹样的线条与色彩完美呈现，赋予每件作品浓厚的民族自信和文化底蕴。更深情诠释了中式浪漫与东方美学，向世界展示了中国传统文化的博大精深和民族自信的坚定力量，开启中式美学的真正复兴。

My husband Yaolong Wu is the head of "XY+Z DESIGN" office and an architect with a strong sense of national mission. With a profound cultural heritage and keen design insight, under his leadership, we have begun to integrate the essence of traditional Chinese architecture into Chinese-style jewelry design, creating a unique and charming new Chinese-style series. He delves deeply into the charm of traditional Chinese culture, perfectly combining ancient architectural patterns with modern furniture and jewelry craftsmanship, inheriting the essence of Eastern aesthetics, and endowing Chinese romance with connotations in a new era.

The elements of traditional Chinese architecture, such as window grilles, cornices and arch of wooden architecture, are reborn and transformed into shining furniture and jewelry. With acrylic, 3D printing, precious metals, and rare jade, we perfectly present the lines and colors of traditional patterns, endowing each piece with strong national confidence and cultural heritage. We try to communicate a more profound interpretation of Chinese romance and Eastern aesthetics, showcasing to the world the vastness and profoundness of traditional Chinese culture and the power of national confidence, opening up the true revival of Chinese aesthetics.

斗拱桌

中国传统再造，以亚克力为主要材质，还原中国斗拱之美。以中国红为主要用色，表达对传统的致敬与热爱。

Bucket Arch Table

With acrylic as the main material, this table is designed to restore the charm of the traditional bucket arch in Chinese wooden architecture. "Chinese red" is used as the main color to express respect and love for tradition.

■

斗拱系列
Bucket Arch Series

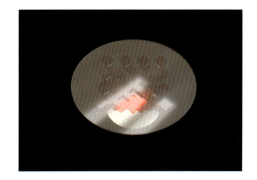

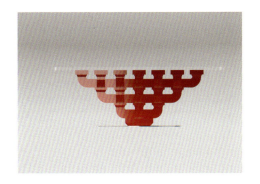
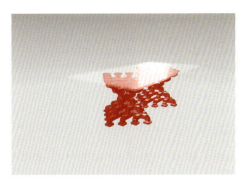
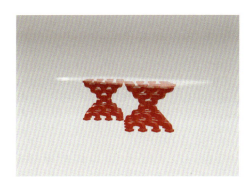

斗拱饰品

中国传统再造，以金、银、铜为主要材质，配红玛瑙、黑玛瑙，共同还原中国斗拱之美。

可拆解设计，深刻了解古典斗拱知识的同时，表达对传统的致敬与热爱。

Bucket Arch

With gold, silver and copper as the main materials, coupled with red and black agate, we interpreted the Chinese tradition by restoring the beauty of bucket arch in China's wooden architecture.

The detachable design is based on a thorough understanding of the knowledge of classical bucket arch in wooden architecture, expressing respect and love for tradition.

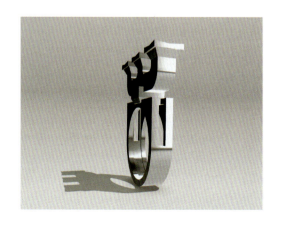

花窗系列
Window Decoration Series

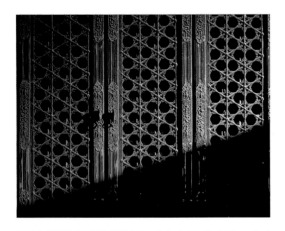

三交六椀菱花窗（图源网络）Three Strips And Six Petals Diamondback window (Figure Source Network)

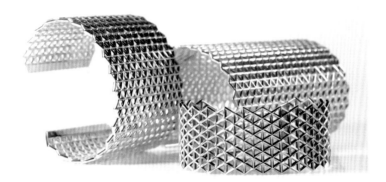

"三交六椀"是装点，也是焦点。

追溯历史的长河，世间美景，皆映照在古窗之外；诗情画意，流淌在一格一扇之间；一窗一景，耐人寻味，是画也是诗。

三交六椀菱花窗一般采取三根棂条交叉于一点，通过一颗金属小钉固定于这个交点，从而在交点处形成一朵六瓣的菱花。这种花纹通常装饰在等级较高的宫殿中，如太和殿、文华殿之类。窗格之于建筑恰似饰品之于美人。

该作品设计用直棂和斜棂相交后组成若干个等边三角形，内涵天地，蕴纳四方，是寓意天地之交而生万物的一种符号。三交六椀天地之间万物生。在光与影的流动中，炫丽珍贵的金属经过现代工艺，映射出故宫的百年辉煌与庄严，也将所有的祈愿和喜悦融入其中。

The "three strips and six petals" pattern is both a decoration and a focal point.

Tracing back the long history, beautiful sceneries of the world are viewed through the ancient windows. The framed sceneries are appreciated as paintings and poems.

The "three strips and six petals" is a pattern for window design. The three strips intersect at a central point, fixed by a metal nail, thus forming a six-petal flower (like a diamond) at the intersection. This pattern is usually used in windows in high-level palaces, such as the Hall of Supreme Harmony and the Hall of Splendor. Windows are to architecture what ornaments are to beauty.

The design of this series consists of several equilateral triangles formed by the intersection of vertical and diagonal strips, which symbolize the intersection of heaven and earth and the creation of all things. In the flow of light and shadow, the dazzling precious metal shines. Modern craftsmanship is used to reflect the century-long glory and solemnity of the Forbidden City, integrating all wishes and joy into it.

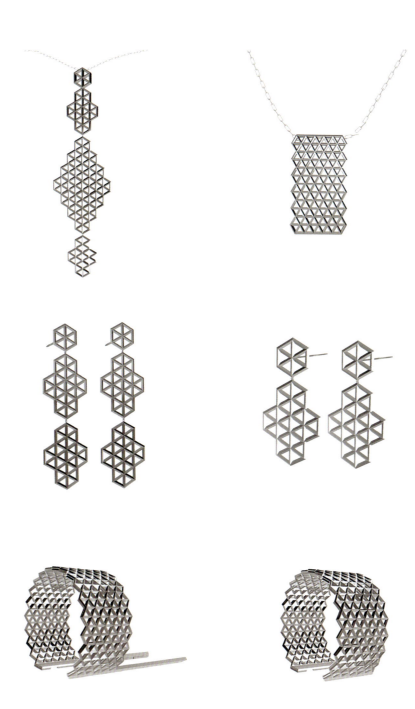

明月海棠 & 繁花方锦

The Floral Series &
The Bright Moon and Begonia Series

繁花系列中经典中式纹样的引入，突破了传统首饰结构，使珠宝始于内、形于外。源于中国古建筑的传统纹样是繁花系列的骨骼。

窗隙流光，繁花似锦。精巧的层叠结构，彰显极简与缜密相融的匠艺，是传统与现代、艺术与技术的统一和重生。对我们而言，如何将窗格中的元素缩小化、精细化，并将其应用至比建筑细小百倍的珠宝首饰上，是我们在产品设计中努力攻克的挑战。

明月海棠系列灵感源于一直深受人们喜爱的"海棠"纹样，幽姿淑态，柔雅圆满，内涵美中之美。它亦有着吉祥的寓意，"棠"通"堂"，寓意着"玉堂富贵""满堂平安"。

重点在于它的多功能性，即是吊坠，又是耳环，还可以是手链。

The use of classic Chinese patterns in the Floral series breaks the traditional jewelry design rules. The traditional patterns originating from ancient Chinese architecture are the "skeleton" of this series.

The exquisite layered structure showcases a craftsmanship that blends minimalism and meticulousness. In the craftsmanship we see the unity and rebirth of tradition and modernity, art and technology. For us, how to refine the elements in the traditional window lattice, and apply them to jewelry that is hundreds of times smaller than buildings, is a challenge we strive to overcome in product design.

The inspiration for the "Bright Moon and Begonia" series comes from the begonia pattern, which has been popular among the Chinese. It is graceful and elegant, with a beautiful cultural connotation. It has auspicious meanings, symbolizing "wealth and prosperity in the jade hall" and "peace at home".

The key feature is its versatility, serving as a pendant, earrings, and even a bracelet.

关于 XY+Z 标志
ABOUT XY+Z LOGO

奇巧兔徽章 Qiqiao Rabbit Badge   银河系列珠宝 Galaxy Series Jewelry

尘世渺小，美的灵魂生而伟大，X、Y、Z 如一串美妙旋律环绕着灵魂。

X、Y、Z 是三维坐标延伸的无限可能性，我们可以在 X 轴、Y 轴、Z 轴里搭建起几何图形，也可以创立无数个空间。

The world is small, while beautiful souls are born great. X, Y and Z surround the soul like a beautiful melody.

X, Y and Z, as three-dimensional coordinates, offer infinite possibilities. We can build geometric shapes with the three axes, and create countless spaces.

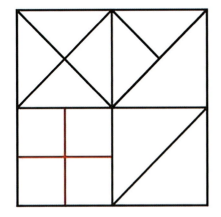
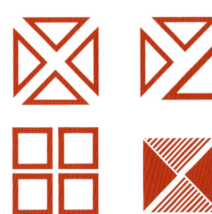

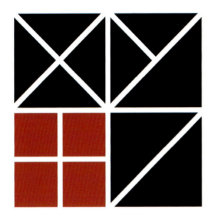

七巧汇星，图案设计灵感取自十二生肖中的元素，并把 XY+Z 标志展开演化成可延万物的七巧板。无穷无尽的几何形态被重新解构诠释，打造出面向八方又向内汇聚的奇巧兔形，寓意迎接八方来客。

The design is inspired by the elements of the twelve zodiac signs, and the XY+Z logo has been expanded and evolved into a seven-piece puzzle that can extend all things. The endless geometric forms have been deconstructed and reinterpreted, creating a strange rabbit shape that faces all directions and converges inward, symbolizing the welcome of visitors from all directions.

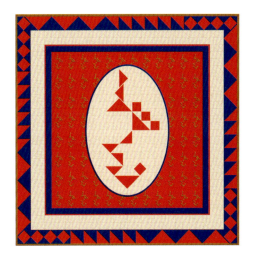

奇巧虎丝巾以"瑞虎"为象征，结合 XY+Z 方格绘出抽象化的虎形，寓意祥瑞，几何图形打破乏味，展现虎的天性与灵气。

The Qiqiao Tiger Scarf features "auspicious tigers" as a symbol. The XY+Z square pattern is used to create the abstract tigers, breaking the boredom of geometrics while showing the nimbleness of tigers.

银河

Galaxy

我们用中国传统的田字格去打造我们的 Logo，更是第四象限法则的灵活运用，代表我们对时间的精准管理。

X、Y、Z，代间表着空间的存在，可以构成经典三元方程，也可以构成未知相加的已知。但，不是设计的全部；X、Y、Z，是三维坐标延伸的无限可能，是坚定不移的设计方向。

符号"+"，是和空间相关的人，是注入空间的情感。

银河系列是以 XY+Z logo 为原型而开展的设计的。

We use the grid pattern (a traditional pattern for Chinese character writing practice) to create our logo. It is also an interpretation of the fourth quadrant rule, representing our precise management of time.

X, Y and Z represent the existence of space, which can form classical ternary equations or something unknown. That's not all of the design. X, Y and Z represent infinite possibilities in extending three-dimensional coordinates and are unwavering design directions.

The symbol "+" refers to people related to space and the emotion injected into space.

The Galaxy series is designed based on the XY+Z logo.

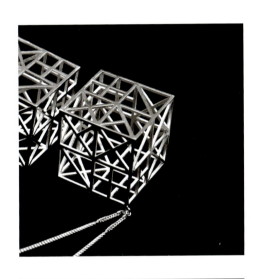

"One Box" have four functions: flower holder, food box, pen holder, and memo.

Flower holder: It would help you snatch a little leisure from a busy life. A bouquet of flowers in it will bring you a serene afternoon in nature.

Food box: The greatest joy of childhood comes from your treasure box. There are so many snacks that have been treasured for a long time inside.

Pen holder: A cluttered desktop won't be good for your emotion at work. This pen holder will give you a clean space and bring your mood in peace.

Memo: Life is filled with hustle and bustle. You can jot down all the trivial things in this memo.

A black acrylic box, a brushed stainless steel lid, a postcard, moon cake box, moon cakes... The "moon cake" is a cute design, showing pure sincerity.

The square black acrylic box is equipped with slots, for stable placement of a stainless steel lid with a strong "XY+Z" style, forming the main body of "One Box".

A customized postcard combines the image of full moon with the "XY+Z" logo, bringing the most sincere blessings of autumn to you on the Mid-Autumn Night. On the opposite side is an instruction of the product. The box is not just designed for the Mid-Autumn Festival, but will accompany you for a lifetime.

■

一盒

One Box

我们为"一盒"设计了四种功能:花器、食盒、笔筒和便签。

花器:忙里偷闲,为劳碌的生活增添一份安宁。一束花,一个清幽的午后,将大自然收录于这一方天地。

食盒:儿时最大的乐趣就是满心欢喜地打开自己的收藏盒,里面还有很多珍藏的小零食。

笔筒:繁杂的桌面会影响工作情绪,这里可以将笔归于井然,也可以将心情归于平静。

便签:生活大量喧哗变装,将自己所有的事情随笔记录,给烦冗放个假。

黑亚克力盒、拉丝不锈钢盖、明信片、月饼小盒、月饼……"月饼"的每一步都经过可爱的设计和制作,塑造我们最纯粹的真诚。

一个四四方方的黑亚克力盒都设有卡槽,将带着"XY+Z"性格的不锈钢盖置入其中,于是便形成了"一盒"的主体。

专门定制的明信片,把"XY+Z"logo融于月圆,在这个中秋之夜,给大家带来最诚挚的秋的祝福;反面则是"一盒"的使用说明,让月饼盒不止留于中秋,而能常伴一生。

XY+Z 激光笔

XY+Z Laser Pen

XY+Z 激光笔，一如既往地多功能。既是激光笔，也是 U 盘。配备红光及绿光，满足不同屏幕需求。黑白两色，诠释经典。

A multi-functional design combining a laser pen and a USB drive all in one. Equipped with red and green light colors to meet different needs. A classic color scheme of black and white is adopted.

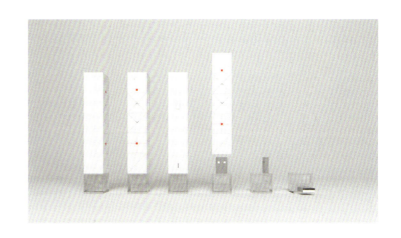

服饰系列

Clothes Series

2

神奇之鹿
THE MAGICAL DEER

蓝鲸
BLUE WHALE

若可的画
PAINTINGS BY RUOKE WU

〔妈妈〕

我的女儿，吴若可。与其说我陪伴她成长，不如说我们在彼此的陪伴中各自成长。这么多年的设计生涯，因为她，我时刻提醒自己设计的初心及纯粹。她从会爬行开始，就喜欢画画。这方面，我始终希望，让她的天马行空得到最自由的释放。

两岁时，她吐字不清地和我说：妈妈，我的身体属于地球，但是我的灵魂，来自星星二号，宇宙只是星星二号的一部分，我的宇宙飞船，隐身在一个丛林里，有一天，我会回星星二号……惊觉不可思议的同时，我开始感慨，或许每一个新生儿原本都来自更高维度的文明，带着高维文明的记忆，随着它们的成长，慢慢地被遗忘。

她热爱设计。从最开始的服装设计，到建筑设计，再到室内设计。常常她的画作，都会给我诸多的灵感。

她热爱和平。无论是核污水排放事件时为海洋感到难过而创作的"蓝鲸"，还是俄乌战争时为期许和平而画的"神奇之鹿"，如果热爱设计更多是受到来自家庭的影响，那么热爱和平，大概就是她来到这个星球的使命吧。

设计需要共情，空间才能有情感。这些年，这个认知我从未动摇——设计师是有使命及社会责任的职业。

来自星星二号的女儿

MY DAUGHTER FROM STAR 2

Ruoke Wu is my daughter. Rather than saying that I accompanied her when she grew up, it's better to say that we grew together in each other's company. Throughout my years of design, I have always reminded myself of my original intention and the purity of design because of her. She loved drawing since she could crawl. In this regard, I have always hoped to give her the most free and unrestrained freedom.

When she was two years old, she couldn't pronounce clearly and one day she said to me: "Mom, my body belongs to Earth, but my soul comes from Star 2. The universe is only a part of Star 2, and my spaceship is hidden in a jungle. One day, I will return to Star 2..." I was very surprised. At the same time, I began to feel that perhaps every newborn came from a civilization in a higher dimensional world. They carry memories in that world, but as they grow, those memories are slowly forgotten.

She has a passion for design, from clothing design in the very beginning, to architectural design, to interior design. Her paintings often give me a lot of inspiration.

She loves peace. She created the "Blue Whale" because she felt sad for the ocean during the nuclear wastewater discharge incident; she painted the "Magic Deer" for peace during the Russia-Ukraine War. If her love for design is influenced by family, then her love for peace is probably part of her mission to come to this planet.

Design requires empathy; only in this way space can have emotions. Over the years, I have never wavered in one thing—designers are occupation with a mission and social responsibility.

/ 星星二号博物馆 /

吴若可诞生于 2014 年 12 月 9 日，拥有设计师家族基因的射手座美少女。

喜欢黑色、冬天、轻音乐，还有妈妈和小盗龙……

热爱自由作画，技能是天马行空的想象力以及日渐展现出人意料的表现力。

梦想成为服装设计师、科学家、建筑设计师、室内设计师，一个伟大的人……

敬畏生命，对世界保持着爱与善意——在家人及宇宙奇迹的支持下，根据若可的意愿设立"奇迹基金"，将其作品出售所得的10% 纳入该基金，用于公益事业，并于每年公布捐赠对象和具体金额。

代表作：神奇之鹿、蚂蚁城、魔法蓝鲸、天使与恶魔。

Ruoke Wu, born on December 9, 2014, is a beautiful Sagittarius girl with designer family genes.

She likes black, winter, light music, as well as Mom and Little Theft Dragon...

She loves to paint freely, and her skills are bold imagination and extraordinary expressiveness.

She dreams of becoming a fashion designer, scientist, architect, interior designer, a great person...

She respects life, holding love and goodwill towards the world. With the support from family and XY+Z DESIGN, she started the "XY+Z Fund". 10% of the earning from the sales of her works will be put into the fund, to be used for public welfare. Names of the beneficiaries and amounts of donations will be announced annually.

Her representative works: Magic Deer, Ant City, Magic Blue Whale, Angels and Demons.

神奇之鹿
THE MAGICAL DEER

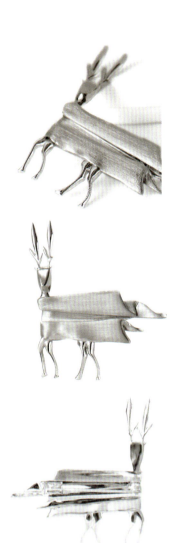

神奇之鹿，时间守护者，周身散发着耀眼的光芒，修长的身姿，清澈的眼眸，灵动的步伐，拥有不死之身，能在永恒的时间之流中自由穿梭，其疾如风，其徐如林，所到之处光芒万丈，黑暗尽除，寓意带来希望与光明。

The magical deer guards the time. She is surrounded by dazzling light. She has a slender body and clear eyes. She runs and jumps nimbly. She is immortal, shuttling back and forth in time. Wherever she goes, there is light shining brightly, and darkness is completely eliminated. She is the symbol of hope and light.

蓝鲸
BLUE WHALE

一个灵魂来自外太空

身体出生于南京的 10 岁小女孩

她热爱生命，天马行空

她不理解核污水为什么入海

不知道被污染了家园的海洋生物将如何生存

她说

蓝鲸都愤怒得跃出海面

于是，她画下了她想象的场景

希望人类能爱护环境，敬畏生命

由此

宇宙奇迹为她制作的公益"蓝鲸"胸针诞生

销售所得的 10% 我们将捐赠于保护动物组织

你的孩子也有很多天马行空

宇宙奇迹愿意为每一位心怀关爱的孩子呈现他的作品

让我们期待更多奇迹的诞生

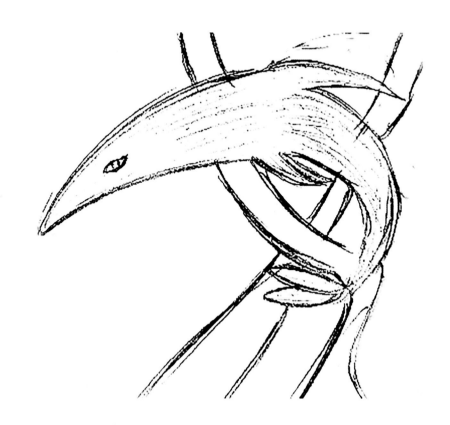

A 10-year-old girl born in Nanjing

with a soul from outer space.

She loves life.

She has a boundless imagination.

She does not understand

why nuclear wastewater is released into the sea.

She does not know

how marine life in their polluted home will survive.

She said that the blue whale is angry and jumps out of the sea.

So, she drew a scene she imagined,

hoping that humans can love the environment and respect life.

"XY+Z" created a "blue whale" brooch for her.

10% of the sales revenue will be donated to animal protection organizations.

Your children also have many boundless wonders in their mind.

"XY+Z" is willing to produce the work of every child

who cares for the earth.

We look forward to the birth of more miracles.

若可的画
PAINTINGS BY RUOKE WU

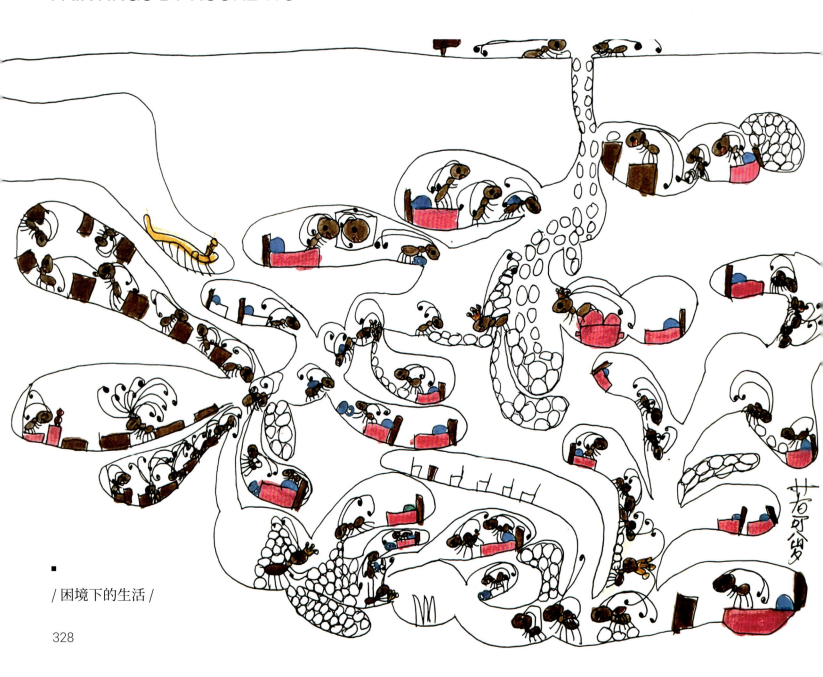

/ 困境下的生活 /

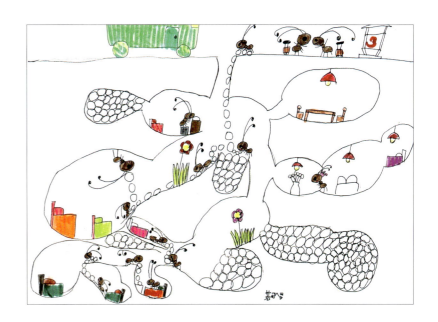

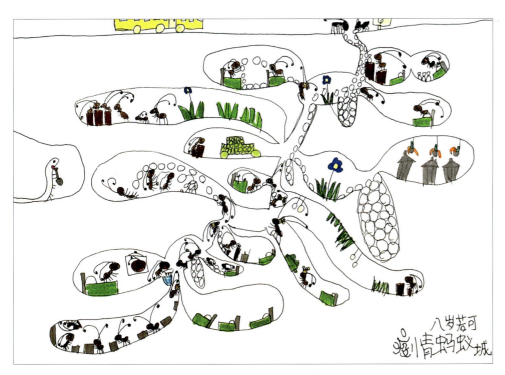

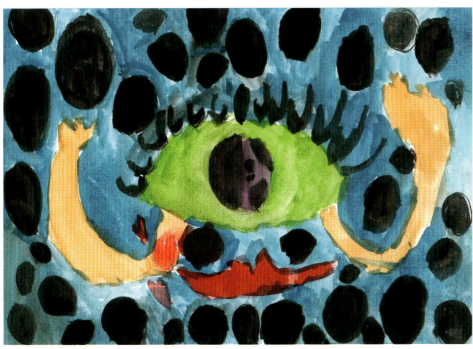

335

后记

AFTERWORD

未来与期许

设计的目的是使静止的文明流动起来，

单向传播的文明多元交互起来，

孤独的文明丰富鲜活起来，

所以需要打破时空、古今、中外、领域的限制，

不同的视角，带来意想不到的惊喜。

未来我们将持续带来，

不同功能与体验完美融合、多元一体的生长型复合空间。

坚持以人为本，关爱环境，致力于创造更美好的生活环境。

继续以独特的复合功能设计，为人们带来前所未有的惊喜与体验。

Future and Expectations

The purpose of design is to make static civilizations flow,

to make unidirectional dissemination of civilizations become diverse interactions,

and to make isolated civilizations enriched and full of vigor.

So we need to break limitations of time and space:

ancient or modern times, domestic or foreign places...

Different perspectives will bring unexpected surprises.

In the future, we will continue to provide

perfect experiences with different functions in composite spaces,

following the principle of people-orientation,

paying attention to the environment,

and committed to creating a better living environment.

We will continue to create unique design

with composite functions that provide new life experience.